The Realness of Things Past

The Realness
of Things Past

Ancient Greece and
Ontological History

GREG ANDERSON

OXFORD
UNIVERSITY PRESS

OXFORD
UNIVERSITY PRESS

Oxford University Press is a department of the University of Oxford. It furthers
the University's objective of excellence in research, scholarship, and education
by publishing worldwide. Oxford is a registered trade mark of Oxford University
Press in the UK and certain other countries.

Published in the United States of America by Oxford University Press
198 Madison Avenue, New York, NY 10016, United States of America.

© Oxford University Press 2018

Library of Congress Cataloging-in-Publication Data
Names: Anderson, Greg, 1962– author.
Title: The realness of things past : ancient Greece and ontological history / Greg Anderson.
Description: New York, NY : Oxford University Press, [2018] |
Includes bibliographical references and index.
Identifiers: LCCN 2018012950 (print) | LCCN 2018022287 (ebook) |
ISBN 9780190886653 (updf) | ISBN 9780190886660 (epub) |
ISBN 9780190886677 (oso) | ISBN 9780190886646 (bb : alk. paper)
Subjects: LCSH: Greece—History—Athenian supremacy, 479–431 B.C.—
Historiography. | History—Philosophy.
Classification: LCC DF211 (ebook) | LCC DF211 .A53 2018 (print) | DDC 938/.04072—dc23
LC record available at https://lccn.loc.gov/2018012950

1 3 5 7 9 8 6 4 2
Printed by Sheridan Books, Inc., United States of America

For Alpana

Contents

PART THREE: *Life in a Cosmic Ecology*

Maps

Ancient Greece

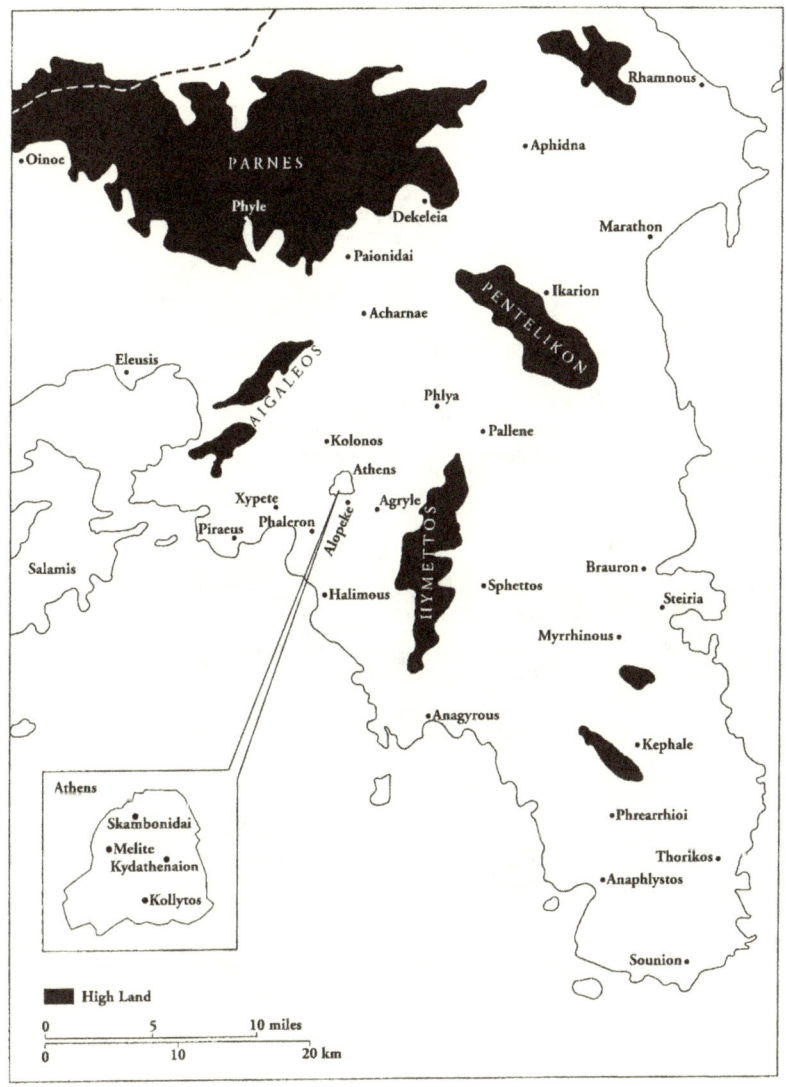

Ancient Attica

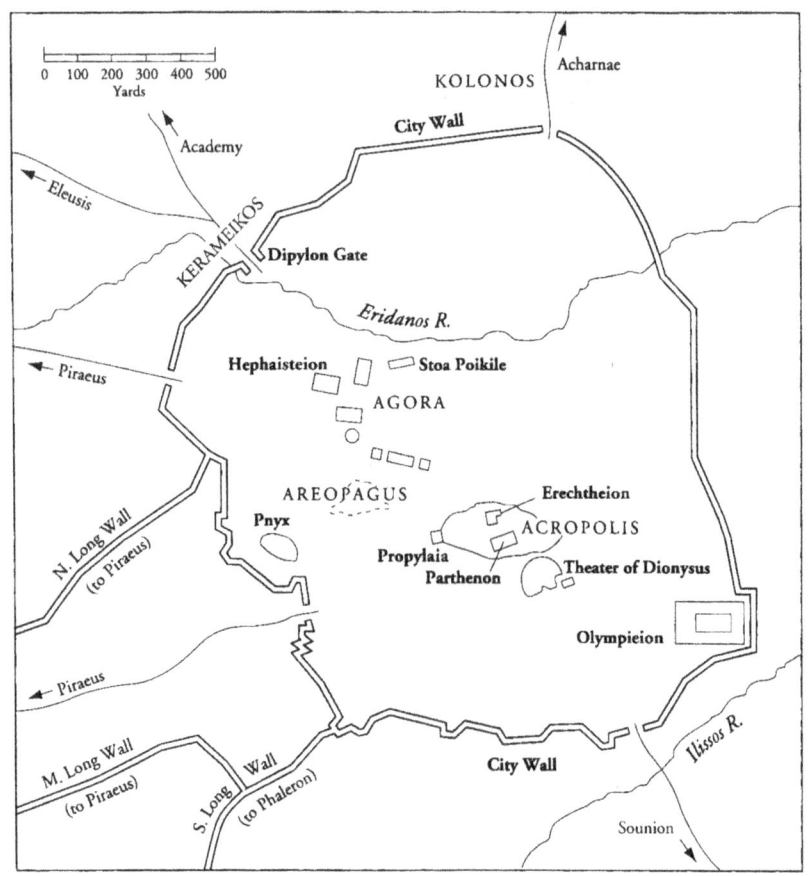

Classical Athens

Preface

SOME TEN YEARS ago, I set out to write a book on the formation of "the classical Greek state," but soon concluded that there was no such "state" there to write about after all. I have been trying to think through the wider historical and philosophical implications of this finding more or less ever since. This very different kind of book is the eventual result.

It is not a work of modest ambition. It proposes a fundamental change in the rules of historical engagement, a paradigm shift that would be very roughly equivalent to the quantum revolution in modern physics. Specifically, it calls for an "ontological turn" in the discipline, a move that would oblige us to account for each past way of life on its own ontological terms. To make the case for this move, the book enlists considerable help from others, drawing liberally on the ideas of various modern theorists and on the experiences of various non-modern peoples, including the classical Athenians, who serve as the work's primary case study. Together, these very different constituencies help us to see how our conventional historicist tools and devices grant us the god-like power to re-engineer the contents of non-modern lifeworlds. They then help us to formulate an alternative historicism, one that will allow us to produce accounts of past experiences that are less ethically questionable, more philosophically robust, and more historically meaningful. And along the way, they help us to explain why there could not have been any such thing as a "state" in the ancient Greek world.

If anybody actually reads this book, it is likely to be controversial. But so long as it helps to generate new conversations that encourage new, less problematic historical practices, it shall have served its primary purpose.

Either way, the work is expressly written to be accessible to a broad, non-specialist, cross-disciplinary readership. While it engages with currents of thought in many different provinces of the intellectual landscape, from anthropology and postcolonial studies to the sociology of science and quantum physics, it strives to avoid the "jargonism," "obscurity," and "self-indulgence" of which more "theoretical" works tend to be accused. Likewise, while the book discusses the classical Athenian way of life in some detail, it assumes no prior specialist knowledge of ancient Greek history or language. The only prerequisites for reading this book are an open mind, a moderate curiosity about the ancient Greeks, and a willingness to reconsider what is arguably the most fundamental historical question of all, namely: What was really "there" at the time?

All Greek language in the book is translated. More often than not, the translations of passages are my own adaptations of standard, published renderings. To ensure the digestibility of a work of such uncommon scope, notes and bibliography have been kept to an absolute minimum. As a general rule, I have cited only those studies which appear to me to be particularly influential and/or representative of wider scholarly currents, thereby providing the most useful points of entry to specialist literatures. And because the study is expressly written to appeal to as many non-specialist readers as possible, the great majority of the works included are Anglophone.

A book about historical conditions of possibility is keenly aware of the particular conditions of its own existence. The debts accrued in this case, both professional and personal, are very considerable.

The Department of History at Ohio State could not have been more supportive, providing me with a warmly congenial professional home, essential material sustenance, and intellectual stimuli of many kinds. For better or worse, it has also allowed me to teach our graduate-level survey of "historical thought" at regular intervals, thereby affording me the chance to nurture and refine the kinds of "big picture" ideas that have now found their way into this book. No less important, the consistent enthusiasm for the project shown by Nate Rosenstein, John Brooke, Dave Staley, and Ying Zhang has been especially energizing, and Ying provided vital specialist help with the section on Ming China in Chapter 6. I thank them all.

Organizers of lectures, seminars, workshops, and conferences have generously invited me to present, allowing me to test the book's ideas on a range of different disciplinary audiences at a variety of venues, including the University of Chicago, University of Michigan, University of Toronto, University of California, Santa Barbara, University of Bordeaux, Northwestern University, Indiana University, Denison University, Wright State University, Penn State University, Rutgers University, and Ohio State. While developing these same ideas, I have also been fortunate to belong to a company of friends and colleagues in ancient Greek studies who meet annually to share works-in-progress. For continually encouraging and critiquing my rather eccentric intellectual endeavors, I am deeply grateful to the group's core members: Ben Akrigg, Ryan Balot, Matt Christ,

Judy Fletcher, Sara Forsdyke, Adriaan Lanni, Jim McGlew, Kurt Raaflaub, Eric Robinson, Bernd Steinbock, Rob Tordoff, Bob Wallace, and Victoria Wohl.

The editor, editorial board, and manuscript readers at the *American Historical Review* also deserve special mention for the guidance they offered when I submitted this book's companion article (Anderson 2015) for their consideration. The journal's famously rigorous review process helped me to tighten up and consolidate the project's core arguments considerably. And it meant a great deal to me that Robert A. Schneider, then editor of *AHR*, showed such faith in my ideas from the beginning. Likewise, my warmest thanks go to Stefan Vranka, the acquisitions editor at Oxford University Press, for believing in the promise of such a "risky" volume in the first place, and to the two anonymous readers for the press, who provided exceptionally thoughtful, helpful, and supportive feedback.

As for more personal debts, James Hanley, my best mate from graduate school, now Chair of History at the University of Winnipeg, always seems to know the right things to say. The good burghers of Yellow Springs, Ohio, could not have been more welcoming to my family since we moved here in 2008. A defiantly unconventional, delightfully twee hamlet, Yellow Springs has been my only true home-from-home in the United States since I first escaped from late Thatcherian Britain to pursue doctoral studies over here. Playing music has been a cherished diversion, and I thank my fellow players in various bands for tolerating my rather modest bass "chops." Few of these bandmates or other friends in the Dayton area have known anything of this book, but all have in some way helped me to write it.

My parents back in England, Sian and Ewan Anderson, continue to give the kind of unstinting, unconditional succour that only parents can provide. And it has been a rare pleasure to be able to share and talk through the ideas of this particular project with my dad, an emeritus professor of Middle East studies. My brother Liam, a professor of political science, is at once a most exacting, gimlet-eyed critic and an inexhaustible fund of fraternal advice and encouragement, emboldening me to face every challenge. I could not be more thankful that he lives so nearby. Closer still to home, life would not have been the same without the society of our various non-human family members, especially Sadie, Oscar, Peanut, and the dear departed Brewster. Our daughters Anna and Lucy have selflessly endured my almost continual state of mental detachment these past years, while still regularly easing the burdens of life and work with the many charms of their company. I owe them both, big time.

Finally, my dear wife Alpana has been there with me on every step of the journey, sustaining me in so many different ways. As a fellow professor, she understands fully the exigencies and demands of academic writing, granting me all the time and space I have needed to complete this project. As a meticulous editor, she has saved me from countless infelicities and gratuitous overstatements. As a specialist in postcolonial studies, she has helped me summon the courage to take my research in bold new directions. And not once has she denied me the chance to talk through the kinds of thorny theoretical issues that a book like this must necessarily confront. She has been a true partner in every sense of the word. The book could not be dedicated to anyone else.

The Realness of Things Past

Introduction

RETRIEVING A LOST WORLD OF THE PAST

E. H. Carr's question "What is History?" needs to be asked
again for our own times. The pressure of pluralism inherent in
the languages and moves of minority histories has resulted in
methodological and epistemological questioning of what the
very business of writing history is all about.

DIPESH CHAKRABARTY

HOW EXACTLY CAN we write true or real histories of lifeworlds whose
standards of truth and realness were quite different from our own? How meaningful is it to speak of states, societies, and economies prevailing in historically
distant environments, where the inhabitants would have found the very idea of a
"state," a "society," or an "economy" entirely meaningless? How can one begin to
make rational, scientific sense of ways of life which presumed the active support
of "supernatural" beings and forces, phenomena that defy our reason and our
science? In short, can our modern academic histories ever hope to speak to the
truth of the realities that were actually lived and experienced by non-modern
peoples?[1]

According to postcolonial theorists like Dipesh Chakrabarty, they cannot. As
Chakrabarty has persuasively argued in his seminal *Provincializing Europe* (2000),
modern academic historicism is itself essentially and inevitably Eurocentrist. Its
fundamental, thought-shaping tools and categories take for granted ways of
knowing human experience that are peculiar to the post-Enlightenment West.
In order to render non-European lifeworlds commensurable with our own, our
historicist devices thus require us to homogenize those lifeworlds by "translating"
them all into Europe's more familiar, more "scientific" terms. They require us
to weave our stories around modern western categories like public and private,
state and society, the sacred and the secular, as if such phenomena were always
potentially there in experience, whether non-Europeans knew it or not. And
in so doing, they occlude or suppress non-western ways of being human, losing

lifeworlds in translation. Accordingly, the histories we write of an "India," a "China," or a "Kenya" will always be mere variations on a master narrative called "the history of Europe." As such, these histories will always end up being stories of incompleteness, inadequacy, and failure.[2]

In the book that follows, I want to build on this powerful postcolonial challenge to conventional historicism and propose an alternative. Along the way, three basic claims are made.

First, our discipline's standard analytical models, categories, and other devices do not just occlude or suppress the experiences of non-western and previously colonized peoples. They require us to flatten and homogenize all non-modern lifeworlds, those of the pre-modern "West" included. The problem with conventional historicism is not its Eurocentrism as such, but its essential, indelible modernism.[3] Second, this modernism is problematic because it shapes the very ontological premises of our practice. The tools of our historicism predispose us to impose post-Enlightenment standards of truth and realness upon all non-modern experiences, western and non-western alike, thereby denying extinct past peoples the power to determine what was and what could be really there at the time. Third, to produce histories that are more ethically defensible, more philosophically robust, and more historically meaningful, we need to analyze each non-modern lifeworld on its own ontological terms, in its own metaphysical environment. To support these three claims, the book uses the proverbially western lifeworld of classical Athens (ca. 480–320 BC) as its primary case study.

As we shall see, the storied "democratic Athens" that one finds in the standard literature is likewise a translated lifeworld, one that presupposes a peculiarly modern way of being human. As such, it is an Athens that is defined by certain absences, inadequacies, and unresolved contradictions. It is an Athens that is always in-the-making, one that is never fully itself.

Thus, in this "democratic Athens," the Athenians' burgeoning sense of "history" somehow coexists with a lingering, strangely uncritical attachment to "myths" about earth-born ancestors, snake-tailed kings, monster-slaying heroes, and sundry other improbabilities. Their commitment to a post-Enlightenment-style "reason" remains continually stifled and compromised by the ubiquitous presence of their irrational "religion." For that matter, their "religion" remains confoundingly entangled with other fields of experience, like state, society, and economy, none of which seem to be neatly bounded and differentiated in the expected rational fashion. In "democratic Athens," the lines between the political and the social, public and private, sacred and secular all as yet remain unhelpfully "blurred."

At the same time, to modern eyes, Athenian "democracy" itself seems to be suspended in a kind of perpetual developmental infancy. Most obviously, it is forever defaulting on its own apparent promise of equality for all, by continually tolerating a manifestly unequal distribution of wealth among native inhabitants, by continually ignoring the flagrant exploitation of innumerable outsiders, like slaves and imperial subjects, and by continually "excluding" females and non-Athenians from the rights and freedoms of "full citizenship." More generally still, the realization of a free, natural individuality, the most basic condition for the existence of any such democracy and citizenship, seems to be forever retarded by the prevalence of non-individualist priorities and imperatives. For some reason, in the progressive, proto-liberal environment of our "democratic Athens," the "traditional" exigencies of corporate well-being, the needs of *polis* and household, still seem to take precedence over the self-realization and self-betterment of the individual subject.

One could go on. But I hope the point is clear enough for now. While the level of theoretical sophistication at work in the field of ancient Greek studies may be significantly higher now than it has ever been before, the universalist protocols, models, and categories of conventional historicist practice still predispose us to produce histories of Athens that are almost inevitably proleptic. They are stories whose contents make full sense only if one already knows how the larger historicist metanarrative will ultimately end. They make full sense only if one already knows that all of the apparent shortcomings of Athenian democracy, citizenship, economics, law, and reason will ultimately be resolved in another, later lifeworld, namely our own.

But what exactly are the alternatives? If conventional histories consistently homogenize and obscure all non-modern ways of being human, western and non-western alike, are there other, less problematic practices available to us, practices which might help us recover something more of the truth of past lives as they were actually experienced at the time?

It might appear so at first sight. For more than thirty years now, historians have been actively trying to access the more subjective realm of "culture," to retrieve the beliefs, ideologies, and discourses through which past peoples apprehended and represented their experiences. But as the book will also show, even the most theoretically refined forms of cultural history still presuppose a conventional historicism. They still authorize us to translate the contents of non-modern lifeworlds, not least because they grant us the final power to decide where subjectivity/culture ends and where objectivity/materiality begins. And in so doing, they predispose us to impose our own peculiarly modern, dualist standards of truth and realness on past experiences, the standards which determine what phenomena

could and could not be really there at the time. Which is to say, the problems
with cultural history, as with all other mainstream historical practices, are not just
methodological or even epistemological. They are ultimately ontological.

It follows then that these ontological issues are the specific problems we need
to address if we are to formulate a genuine alternative to our conventional historicism. And this means that we need to rethink the very philosophical foundations
of our discipline, untethering them from modernity's dualist certainties. As
the epigraph on the first page suggests, we need to find a new and significantly
different answer to E. H. Carr's question: "What is History?" Specifically, we
need to imagine and act upon an alternative mode of historicism, one that
encourages us to transcribe rather than translate the contents of past experiences.
We need a historicism that allows us to understand each non-modern lifeworld
more on its own ontological terms, in its own metaphysical conjuncture. We need
a historicism that allows us to recognize that Olympian gods were just as self-evidently real to the Athenians as a free market economy is to ourselves. In short,
I propose, we need to move beyond cultural history and take an ontological turn
in our practice.[4]

The book is divided into three parts. Part I lays out the central problem,
specifying the kinds of difficulties that arise when we use conventional modern
devices to historicize a non-modern lifeworld, in this case the *polis* of the classical
Athenians. It begins by introducing "democratic Athens," the standard scholarly
account of this *polis*, which quite explicitly uses a modern, universalist model
or template of social being to historicize the *demokratia* of the Athenians, their
politeia or "way of life" (Chapter 1). This account is problematic for a number
of reasons. It presents a world that is riddled with improbable contradictions
(Chapter 2). It depends on master categories which have no correspondence
to phenomena that were known and experienced by the Athenians themselves
(Chapter 3). And it encourages us to tell questionable stories about the formation of their *politeia*, narratives which always end up figuring Athens as an imperfect anticipation of a modern lifeworld, never as a fully realized version of itself
(Chapter 4). The primary cause of all these problems is our universalist template
of social being, which is taken for granted by all mainstream forms of historical practice, materialist and culturalist alike. For this model obliges us to homogenize the past's many different ways of being human at the ontological level
(Chapter 5).

Part II formulates the book's suggested solution to its central problem.
Elaborating on recent arguments made by influential anthropologists, it builds
ethical and philosophical cases for an ontological turn in the study of non-
modern experiences. To help frame the ethical case, the book first surveys

ontological variabilities across time and space using evidence from a wide range of non-modern environments, including Ming China, precolonial Mexico, medieval Europe, and indigenous lifeworlds in Amazonia and South East Asia (Chapter 6). It then looks in some detail at the core metaphysical and ontological commitments which sustain our own modern capitalist way of life, our mainstream sciences, and thus our conventional historicist practice, showing quite precisely how these commitments predispose us to produce ethically questionable histories, accounts which effectively modernize all non-modern lifeworlds at the ontological level (Chapter 7). Next, the book presents its philosophical case, enlisting the help of numerous prominent authorities in a wide array of fields, from posthumanist studies to quantum physics. After directly challenging modernity's dualist orthodoxies from a variety of different critical perspectives, it goes on to propose an alternative, non-dualist metaphysics, a counter-metaphysics that would authorize us to historicize each past way of life in its own distinct world of experience (Chapter 8). Part II duly concludes by discussing what this ontological turn would involve in actual practice (Chapter 9).

By way of illustration, the third and last part of the study revisits the primary case study, showing how the proposed paradigm shift would help us to produce a new, more historically meaningful account of the Athenian *politeia*. It begins by reconstructing the a priori template of social being upon which *demokratia* was premised. This model seems to have taken the form of a kind of cosmic ecology of gods, land, and *demos* ("people") in Attica, a kind of perpetual symbiosis between the human and non-human constituents of the *polis* (Chapter 10). The principal practices and mechanisms of the Athenian *politeia* are then re-examined in their original metaphysical conjuncture, the result being a radically different account of an ancient way of life.

For as we shall see, an ontological turn fundamentally alters our sense of the significance of ritual engagements with gods (Chapter 11). It prompts us to rethink the very nature of the *polis* as an entity and the contributions made by women and households (*oikoi*) to its vitality (Chapter 12). It changes the way we view the various societal functions performed by demes, phratries, tribes, and other autonomous corporations in Attica (Chapter 13). And it decisively changes how we make sense of the quiddity, the agency, and the governmental competence of *demos* itself, the human subject of *demokratia* (Chapter 14). Furthermore, when it comes to understanding the flows of resources that were necessary to sustain this ancient way of life, an ontological turn also provides us with an altogether more productive alternative to conventional, modernist economic analysis, one that highlights the crucial ecological roles played by gods, females, elites, and slaves (Chapter 15). A number of possible objections to the book's novel account of the

Athenian *politeia* are then anticipated and addressed (Chapter 16). A brief conclusion ends the study by considering the broader intellectual implications of the proposed paradigm shift, both for disciplines which expressly concern themselves with the analysis of non-modern lifeworlds and for contemporary critical theory.

Let us now turn to the book's primary case study and see how our conventional historicism requires us to translate the *politeia* of the classical Athenians.

PART ONE

Losing Athens in Translation

*The moment we think of the world as disenchanted . . . we set
limits to the ways the past can be narrated.*

DIPESH CHAKRABARTY, Provincializing Europe

I

Our Athenian Yesterdays

THE AREA AROUND the Athenian acropolis has been continually settled since the Neolithic Age (ca. 7000–3000 BC). At some point during the subsequent Bronze Age (ca. 3000–1100 BC), perhaps shortly before 2000 BC, the course of Athenian history was then irrevocably altered by the arrival of an Indo-European population, namely "the Greeks," who brought with them a new language, new practices, and new gods. During the Late Bronze Age (ca. 1600–1100 BC), these immigrants went on to establish a complex, hierarchical form of social organization, similar to the better attested examples at sites like Mycenae and Pylos, whereby a monarchic authority directly managed the production and distribution of essential resources for the entire surrounding region from a palace on the acropolis itself. But by the Early Iron or "Dark" Age (ca. 1100–700 BC), this Near-Eastern style "Mycenean" society had collapsed in mysterious circumstances, and habitation was now reduced to a cluster of small villages around the former citadel, with relatively modest levels of social stratification. Only in the eighth century did these villages begin to coalesce to form the *polis* of Athens. And after several centuries of elite domination of early *polis* government in the archaic era (ca. 700–480 BC), it was only in 508/7 BC, with the watershed reforms of Cleisthenes, that the foundations were finally laid for the democratic *politeia* that would prevail in the classical era (ca. 480–320).

This, at any rate, is the master narrative that one usually encounters in our most recent textbooks, the storyline that we commonly pass on to our students.[1] And no doubt there is a certain kind of truth to this story. But as all Greek historians would readily concede, this is not a truth that the classical Athenians themselves would have recognized. They knew nothing of the many dramatic shifts, ruptures, and transformations which punctuate our modern accounts. They knew nothing of any Indo-European immigrants, Bronze Age "redistributive economy," or Dark Age. They knew nothing of any grand evolutionary trajectories from *Gemeinschaft* to *Gesellschaft*, from palace to *polis*, from "oriental" monarchy to a more familiar, more progressive, more "western" mode of order. Indeed, all evidence suggests that they barely even remembered the reforms of Cleisthenes, those landmark "democratic" measures which loom so large on our own historical horizon. Instead, they tended to see a past of more or less seamless continuity, one that stretched all the way back to the time of the first kings, Cecrops and Erechtheus, who were born literally from the soil of Attica. As far as they were concerned, their *polis* had been there more or less from the start. And their *politeia*, the distinctive "way of life" that they called *demokratia*, had prevailed essentially unchanged since the time of Theseus, the one-time Minotaur slayer and *polis* founding father par excellence.[2]

Yet when we moderns then try to make sense of that *politeia*, we tend to analyze it as if the Athenians were living out our modern vision of their history, not their own. Instead of asking how the authors of that *politeia* objectified and explained their way of life, we try to make sense of it using our own modern social scientific models and categories, as if we know better. Let us now look in a little more detail at how this conventional approach works out in practice.

A Universalist Template of Social Being

Modern attempts to analyze and explain the Athenian *politeia* tend to order the evidence of ancient experience in a very particular way. Invariably, they involve mapping this evidence onto a kind of universal analytical template, a social scientific model that historians rarely if ever see the need to defend or justify per se. Presumably, this is because the model in question seems to capture a timeless, objective truth of social being, one that will hold for all complex worlds of the past.

This universalist model begins by imposing a primordial dichotomy on all the contents of experience, separating once and for all what seem to be objectively real, material phenomena, like practices and institutions, from subjective, cultural phenomena, like beliefs, values, and ideologies. It would then assign each of these phenomena to its designated place in "the world," which the model objectifies in abstract, spatial terms as a kind of disenchanted, functionally divisible arena or terrain. Among the various possible divides in this terrain, the most fundamental would be that which separates a human, societal order from a non-human order of nature. And within the broad confines of the human order, experience can then be further subdivided into various realms, spheres, and fields, like the sacred and the secular, the public and the private, and the political, the social, and the economic. As for the human inhabitants of this abstract societal terrain, the model presumes them all to be natural, self-actualizing, psycho-physical individuals, albeit individuals whose thoughts and actions are influenced by whatever cultural phenomena happen to prevail at any given time.

Like our counterparts in most if not all other fields of history, we specialists in ancient Greece take this abstract spatial model of social being largely for granted. Though we might quibble over how precisely its constituent categories ought to be applied, we rarely acknowledge its status a "model" at all. Either way, the accounts of the Athenian *politeia* that one finds in mainstream scholarship are almost invariably mapped onto this same basic theoretical template. In the remainder of the chapter, I will try to summarize what the current consensus account of this *politeia* then looks like, noting only the most significant points of debate and disagreement along the way. At the same time, this summary account can serve as a general introduction to the salient features of the Athenian way of life for those who are less acquainted with the subject.[3]

The Primacy of "the Political"

The consensus account of Athenian social being would begin with the secular, public field of the political, since it is generally assumed for analytical purposes that a Greek *polis* was an essentially secular, political entity, one that was ultimately defined by its mode of government. Many today, one suspects, would agree with the following judgment of Christian Meier, one of classical antiquity's most distinguished historians:

> The political was . . . the central element in the life of Greek society, and of Athenian society in particular. . . . [It was] at the center of all perceptible change, and especially perceived change. The only change that men could perceive was the sum of political actions and events, since everything outside politics was essentially static, or changing only slowly and for the better. . . . [E]verything was determined by political action.[4]

Accordingly, we tend to see the essence of the classical Athenian *polis* as its *demokratia*, which we take to be a political regime or system that bore a general family resemblance to our own liberal governments of today. For like modern liberal democracy, Athenian *demokratia* appears to us to be a distinctly egalitarian, individualist system of rule. Like our own form of government, it seems to have been expressly designed to allow each and every "citizen," each member of the "people" (*demos*) in question, to exercise something like an individual right to political self-determination.[5]

Of course, we know that this ancient "democracy" was suspended on two occasions in the later fifth century, when the oligarchic regimes of "the Four Hundred" (411 BC) and "the Thirty" (404–403 BC) briefly took power. But both of these developments were clearly aberrations, caused by the extraordinary

stresses of the Peloponnesian War (431–404). More important, we also recognize that membership of the Athenian citizen body was denied to a rather large
portion of the 200,000 or more individuals who actually inhabited Attica at any
given time, being limited as a rule to the region's free, native-born, adult males.
But these exclusions should not obscure the fact that even relatively poor, uneducated males were able to enjoy the privileges of full citizenship in the classical era.
This, we assume, is what the Athenians meant when they said that in Athens one
was "free" to "live as one likes."[6]

To become a citizen, an Athenian had to enroll at age eighteen as a constituent
of one of over a hundred "demes," local administrative units that were scattered
across the region. In so doing, one simultaneously became a member of one of
Attica's ten tribes (*phulai*), the basis for participation in much of the organizational life of the *polis*, each one of which contained an admixture of demesmen
from the three principal sub-regions of Attica—the city, the coast, and the hinterland. And enrollment as a citizen also entitled one to perform a wide variety of
public functions as an individual.[7]

Most important of all, every one of the ca. 30,000–40,000 full citizens in
Attica was equally entitled to attend, address, and vote in the meetings of the assembly (*ekklesia*), which were held outdoors on the Pnyx hill on just forty or so
days per year. This institution was charged with producing binding resolutions on
most if not all matters of Athenian policy and legislation, thus serving in effect as
the sovereign body in the *polis*. In so doing, it exemplified what we see as the egalitarian, democratic principles of government in Athens. For in the assembly, there
were no presidents, prime ministers, cabinet members, or formally empowered
political leaders of any kind. All outcomes were determined by a simple majority
vote of those present, with somewhere between around 4,000 and 6,000 citizens
usually attending on any given occasion. Unlike the representation-based systems
of liberal modernity, the Athenian government was thus what we usually call a
direct democracy.[8]

To bring some necessary structure to assembly meetings, another institution,
the council of 500 (*boule*), was tasked with setting the agenda of motions for discussion, serving in effect as a kind of standing committee for the sovereign body.
The council too was run according to what look to us like democratic, egalitarian
logics and principles. To ensure that all parts of Attica were represented in its
deliberations, each tribe had to provide 50 volunteers to fill its ranks annually,
with quotas ensuring that these contingents included delegates from every district of the region. And since no one could serve more than two non-consecutive
terms on the *boule*, its membership changed completely from one year to the
next. At the same time, the council served effectively as the permanent or continuous "face" of Athenian government, since at least some of its members had

to be on duty at all times, ready to receive the approaches of all Athenians and non-Athenians who wished to have items of business placed before the sovereign *ekklesia*. If approved by the council, these matters would be presented at the next assembly meeting as "preliminary motions" (*probouleumata*), then discussed and voted on by the citizens present, with all final "resolutions" (*psephismata*) later inscribed on stone as public records.[9]

As for the actual deliberative process in the assembly, this would often be guided by "speech-makers" (*rhetores*), individuals who volunteered to "advise" the *demos* by proposing specific lines of action. More often than not, these "advisors" came from a tiny, super-wealthy minority of the population, roughly the wealthiest 1–2% of Athenian families, whose members alone possessed the leisure, education, speaking ability, general knowledge, and international connections to serve in this role as de facto political leaders. While these elite speakers never formed a common interest group among themselves, figures like Themistocles, Cimon, Pericles, Alcibiades, and Demosthenes were able to exercise a significant personal influence over Athenian policy-making for many years.[10]

According to modern conventional wisdom, the legal system in classical Athens was, if anything, even more democratically egalitarian than the political institutions. Again, there were no trained experts or professional authorities of any kind on hand to direct and manage the system. Instead, all functions were again performed by civilian volunteers. While the Athenians had a number of special courts for dealing with certain unusually problematic cases, particularly those which might cause the "pollution" of the *polis* by bloodshed, the vast majority of proceedings were handled in the regular jury courts (*dikasteria*). Here, all verdicts were determined by a majority vote of the civilian jurors (*dikastai*), who were chosen from an annual pool of 6,000 volunteers and served on panels that usually ranged in size from 200 to 1,500 members.[11]

Most routine court cases fell into one of two categories. Private injury suits (*dikai*) could only be brought by the family of the injured party. Public interest cases (*graphai*), where the accused was charged with inflicting some kind of harm upon the *polis* as a whole, could be brought by any full citizen. But in both instances, it was still left to individual civilians to consult the pertinent laws, initiate the proceedings, gather the necessary evidence and witnesses, prosecute the case in court, and sometimes even impose the designated penalty. In practice, *graphai* tended to be initiated only by wealthy elites, since a heavy fine of 1,000 drachmas, roughly three times the annual income of a poor Athenian, was imposed on prosecutors who failed to muster one-fifth of the jury vote. But in theory, any full citizen, no matter how humble, could bring such a case against any other.[12]

To enforce the decisions of courts and assembly and maintain the daily well-being of the *polis*, the Athenians relied on the council of 500 and a small army of office-holders, most notably the nine "archons," the pre-democratic "rulers" of the *polis*, who now served largely as legal and ceremonial functionaries. As a general rule, any Athenian citizen could perform any of these official roles, many of which were supported by the services of a large, if indeterminate, number of publicly owned slaves. Typically, magistrates held their positions for just a single year, discharging their duties as members of a panel or board, not as individuals. Their multifarious assigned tasks ranged from the management of treasuries, roads, and water supplies to the supervision of festivals, markets, legal processes, building regulations, dung-collection, and detainees in the city's small jail.[13] But they conspicuously did not include what we would call police functions. While for a time the Athenians used a force of several hundred specially imported, publicly owned archer-slaves from Scythia to control large crowds, they otherwise left citizens free to police themselves, using available legal mechanisms if necessary.[14]

As already indicated, all of the seven hundred or so magistratical positions were filled each year by lay volunteers. In a small number of cases, including those of general and treasurer, where effective performance required a significant level of knowledge and/or experience, appointees were chosen by election and came invariably from the wealthy, well-educated minority. But by the middle of the classical era, the vast majority of poliadic officials were chosen by lottery and paid a small stipend for their services, thereby extending what we see as the principles of democratic government even to the appointment of functionaries.

Finally, we would also include in the political field the various mechanisms the Athenians used to organize military forces. The principal mechanisms for levying land forces were again the ten tribes, each one of which was expected to furnish a squadron of cavalry and a regiment of "hoplites" (heavy infantrymen). Originally, each tribe was also required to provide a field commander (*strategos*), though at some point in the classical era, these ten generals came to be elected from the citizen body as a whole.[15]

As for the specific roles played by individuals in war-making and the defense of the *polis*, these were determined primarily by one's personal resources, since it was generally left to citizens to train themselves and furnish their own equipment. Hence, the cavalry tended to be manned by the tiny wealthy minority who could afford to own or lease horses. And anyone who could afford a shield, sword, breastplate and the like was eligible to serve as a hoplite, while the less prosperous could volunteer for roles as lighter-armed *psiloi* or "peltasts." Meanwhile, the thousands of oarsmen required to man the triremes of the Athenian navy, the basis of Athenian military power in the classical era, generally came from the ranks of poorer citizens, resident aliens, citizens of "allied" *poleis*, and slaves. Here too, the less affluent were induced to serve by the payment of a small stipend. Commanding all these forces were the ten generals, who typically came from the same wealthy minority that provided the de facto political leaders, quite often serving in both capacities simultaneously.[16]

A Society of Households

If modern consensus opinion thus regards the Athenian *polis* as an essentially political entity, one defined by the democratic public institutions of a secular political field, it would then counterpose to this a similarly discrete, similarly secular, but private field of the social. It is true that in recent years some have questioned the application of these particular categorical distinctions to the Athenian context. Given a world where the government was run entirely by civilians, a growing number would prefer to think of the Athenian *polis* as a kind of fusion of state and society, where the lines between political and social, public and private, were somewhat blurred.[17] But in practice, as this formulation indicates, all specialists happily continue to use all of the categories in question, regardless of how well they correspond to the ancient evidence.

As for the contents of the putative social field in Athens, all would likewise agree that its definitive essence was the household unit or *oikos*. Unlike a modern nuclear family, this ancient Greek *oikos* was a complex of persons, property, and other resources, one that provided the basic means of existence for all its members. Ideally self-sufficient, it was typically a multi-generational, patrilineal social unit. Under normal circumstances, the titular head (*kurios*) and public face of an *oikos* was the father figure, who inherited his property and resources from his paternal ancestors, thereby perpetuating their lineage. Athenian wives thus married into existing family units from outside rather than forming new ones with their husbands.[18]

Marriages in Athens did not have to be formally licensed or sanctioned by public authorities. So long as two Athenians were seen to be committed to "living together" (*sunoikein*) as husband and wife, their relationship was deemed to be legitimate in the eyes of the *polis* community. Marriages were generally initiated and arranged by the parents of the two parties, usually when the bride was only in her mid-teens, so as to maximize her chances of producing suitable heirs thereafter. And to help pay for her upkeep after transferring from one *oikos* to another, her birth family might provide a dowry.[19]

According to the standard views, female experience after marriage was defined largely by negatives. Athenian women could not participate in political life, could not represent themselves in court, could not usually own property, and, unless they were poor and had no choice, could not work outside the house.

Most women were uneducated by modern standards, learning as girls little more than was necessary to perform their primary role in life, which was to serve the households of their husbands. This meant above all bearing and raising children, weaving, directing slaves if they had them, and cooking, cleaning, and other menial tasks if they did not. Contact with men from other families was limited and controlled, and women from wealthier Athenian families seldom appeared in public. The principal exceptions to this rule were funerals and festivals, where women sometimes assumed prominent roles. In certain cases, like the Thesmophoria, an important annual fertility festival of Demeter, the celebrants were exclusively female.[20]

Also conventionally assigned to Athenian society would be networks of friends and neighbors, along with a wide variety of more formally constituted associations, whether based upon ties of kinship or locality. Moderns commonly see these various associations as comprising a kind of "civil society" in Attica, occupying a space somewhere between the private life of the *oikos* and the public life of the *polis*. Under this broad category heading, we would include a range of venerable local cultic organizations like the Marathonian Tetrapolis (Marathon, Probalinthos, Oenoe, Trikorynthos) and the Tetrakomoi (Piraeus, Phaleron, Thymaitadai, Xypete). Athenian civil society would also include "religious clubs," like the *orgeones* and *thiasoi*, along with the extended kinship groups known as *gene* ("clans"), whose members were all claimed descent from some common, somehow illustrious ancestor. It would likewise include the *phratriai*, more diffuse descent groups to which most if not all full Athenian citizens belonged. It would include the 139 demes, each one of which was a kind of neighborhood-level *polis*-in-miniature, with its own volunteer officers, its own local decrees, and its own calendar of religious festivals. And it would include the ten tribes, each of which likewise had its own cohort of voluntary officials and convened regularly for sacrifices at the tomb of its namesake hero, a figure from Athenian legend who was deemed to be its "founding father" (*arkhegetes*).[21]

Finally, our standard accounts of Athenian society would also include the experiences of the many non-Athenian inhabitants of Attica. Growing numbers of resident aliens or "metics" (*metoikoi*) moved to the region over the course of the classical era, with perhaps as many as 20,000 settled there by the fourth century. So long as these persons were endorsed by an Athenian sponsor (*prostates*), officially registered as metics, and paid a modest tax of one drachma a month, they could share in a number of the rights and obligations of citizenship, including the right to legal representation and the duty to fight for the *polis* when necessary. However, as we conventionally see them, metics remained at best only partial citizens, since they were also denied the rights to own property in Attica, to speak in the assembly, or to hold political office of any kind.[22]

Slaves, by contrast, had no rights in Attica to speak of at all. Evidence for the total numbers of slaves in Athens is notoriously problematic, with estimates varying widely from around 50,000 to as many as 150,000 at any given time in the classical era. Most scholars would probably accept a relatively conservative number in the region of 80,000. Either way, the vast majority of these individuals were non-Greeks, imported through slave markets from regions like Anatolia, Thrace, and Scythia. While all slaves were denied any basic sense of what we would call personhood, their life conditions varied considerably, depending largely upon the kinds of labor they were required to perform. In all likelihood, the harshest conditions were experienced by those who worked on the estates of the wealthy, in mines, in the larger workshops, or in the city's brothels. As a general rule, life as a domestic worker or servant was slightly more comfortable, since it made little sense for Athenian households to mistreat their human assets. Otherwise, a substantial number of slaves were also purchased to serve the needs of the *polis* as a whole. These public slaves (*demosioi*) performed various, sometimes skilled roles for the *demos*, their tasks ranging from record-keeping and coin-testing to dung-collection and crowd control. But little is known of their everyday circumstances. Whatever the case, while it is known that some slaves were granted a certain independence or even freed by their owners, these situations seem to have been exceptional. In the rare instances when slaves were freed, they were permitted to register as metics.[23]

An Ancient Economy

Talk of slavery brings us to the other more or less private, more or less secular field or system of Athenian experience, which would be the economic. Here again one should acknowledge that there is some disagreement among specialists about the precise valence of this category in ancient experience. This after all was a world where there were no capitalist-style limited liability corporations or "markets" for free labor, a world where one typically depended on one's own household for one's livelihood. But all specialists still continue to speak of an "economy" in Athens, as if such an object were really there at the time, whether the ancients knew it or not.[24]

As for the phenomenal contents of this field, consensus modern opinion would see a dominant agrarian component in the Athenian economy, with agriculture remaining the principal source of livelihood for the roughly 50% of Athenian households that were located in and around the settlements of rural Attica.[25] Since the classical Athenians chose to depend increasingly on imported grain, many Attic farmers are thought to have shifted production during this time towards potentially more lucrative commodities, like olives, grapes, and dairy goods. Private holdings ranged from small family-worked plots to large estates that required the labor of teams of slaves. As for other forms of primary production, the ample silver reserves of southern Attica, like the stone quarries on mounts Pentelikon and Hymettos, were publicly owned resources. But substantial private profits could be made by those who were wealthy enough to purchase mining rights from the *polis* and lease the requisite numbers of slaves to exploit the silver deposits.[26]

By contrast, urban economic life in Athens and Piraeus was extremely diverse, with as many as 170 different occupations attested. In the port and markets, for example, one could find merchants, foreign traders, bankers, insurance brokers, and all manner of retailers. Others offered services of many kinds, from teachers, doctors, and nurses to builders, tanners, fullers, cobblers, musicians, innkeepers, and prostitutes. Still others supported themselves by various forms of manufacturing, such as pottery, carpentry, stonemasonry, and metal-working, with the size of their operations again determined primarily by the number of slaves they possessed, since citizens did not typically work for one another. And a large number of public functionaries were duly required to monitor and, if necessary, regulate all this economic activity, including market managers (*agoranomoi*), grain custodians (*sitophulakes*), coin testers (*dokimastai*), and controllers of weights and measures (*metronomoi*).[27]

As most modern observers would acknowledge, this relatively complex division of labor in Athens produced a conspicuously uneven distribution of wealth among *oikoi*. While the Athenians did not keep systematic statistical records of such matters, the general impression is that a majority of households, possibly as many as 60%, were considered relatively "poor," functioning around or just a little above subsistence level. Sometimes referred to collectively as the "thete class," these poorer Athenians would have typically served as naval oarsmen and light-armed troops in the military campaigns of the *polis*. Above them, perhaps as many as 30% of all families, the "hoplite class," were moderately prosperous, while no more than 10% were deemed to be self-evidently "rich." Among the latter, we should also distinguish a small minority, perhaps 200–400 families at most, who were significantly more affluent still, possessing resources that were a hundred or more times greater than those held by the average poor household. Though again it is hard to be precise about such matters, we can be fairly confident that slave-holding tended to be concentrated in the wealthier segments of the population. If we accept the rough figure of 80,000 total slaves in Attica, and thus an average of two per household, many poorer families will have owned no slaves at all, since some super-wealthy *oikoi* are known to have possessed well over one hundred.[28]

Under normal circumstances, public revenues were relatively modest. The period of the Athenian empire (ca. 454–404 BC), when an annual tribute was exacted from as many as 170 other *poleis*, was conspicuously exceptional. There was no routinized system of annual taxation in Athens itself, though ad hoc "contributions" (*eisphorai*) could be demanded of wealthier Athenians when some pressing need arose. Regular revenue sources included leases of mining rights, rents from sacred lands, import duties, tolls, and law court fines. The precise details of Athenian financial administration remain elusive, but it seems that four different treasuries were used in the classical era for receiving and disbursing a variety of funds. Two, the treasury of Athena and the treasury of the Other Gods, were used to cover general, everyday expenses, especially those incurred by the ritual activities of the *polis*. The nature and purpose of the third, the *demosion* ("public treasury"), remain somewhat unclear. The fourth, the treasury administered by the Hellenotamiai ("treasurers of the Greeks"), was originally established to support the activities of the Athenian-led Delian League just after the Persian Wars (480–479 BC). But as Athenian control over the league began to assume more of the character of an empire, this fund was transferred to Athens in 454/3 and later amalgamated with the *demosion* around 415.[29]

Finally, under the rubric of economy one should mention the phenomenon of "liturgies" (*leitourgiai*) or public works. These were ostensibly voluntary contributions to the general well-being which the *demos* regularly required of its most prosperous members. Among the more important of these liturgies were the "trierarchies," the maintenance of each of the fleet's 200 or more trireme warships, the sponsorship of choruses at festivals, and sundry other ritual and ceremonial outlays. In some of these undertakings, especially the trierarchies, the expense involved could be very considerable, up to ten or more times the annual income of a poorer Athenian household.[30]

Polis *Religion*

With the mention of festivals and the ritual activities of the *polis*, we now pass from what mainstream accounts would consider the secular fields of experience to the sacred realm of religion. All specialists would recognize the significance of ritual to the ancient Athenian way of life. After all, participation in religious practices was evidently routine and frequent for all members of society, metics and slaves included. And this participation came in a wide variety of forms, which again are usually grouped under the headings public and private.[31]

For example, as private individuals, Athenians could, say, consult oracles, seek initiation into the Eleusinian Mysteries, or make healing offerings for themselves or others. Families ritually celebrated the arrival of new members, like children, brides, and slaves, made offerings at the graves of their ancestors, and maintained cults of their own household gods, like Zeus Herkeios and Apollo Patroos. And as already mentioned, ritual formed the primary group activity for a range of different groups and associations, such as the phratries, the *gene*, and the Marathonian Tetrapolis. The demes likewise had their own calendars of regular festivals and sacrifices in Attic localities, while members of the ten tribes convened regularly for shared rites at the tombs of their respective eponymous heroes. At the same time, dozens of larger, more inclusive ritual events, like the Panathenaia, City Dionysia, and Anthesteria festivals, were staged in Attica on specified days each year. Publicly funded and centrally administered, these events provided the backbeat for the regular rhythms of life across the *polis* as a whole.[32]

Indeed, it is commonly said that religion was everywhere in the *polis*. The traffic of offerings to gods and other superhumans was more or less constant and ubiquitous, taking place at sites which ranged from tiny household altars and roadside shrines, to graves, caves, and hilltops, to grand sanctuaries like those of Demeter at Eleusis, Poseidon at Sounion, and Athena Polias, the patron of the city, on the acropolis citadel itself. Moreover, some form of prayer, oath, sacrifice, or other ritual act accompanied almost every significant human undertaking, from major life events, like conception, birth, and death, to citizenship enrollments, political deliberations, law court proceedings, commercial transactions, overseas voyages, and battlefield preparations. Somehow, it seems, the Athenians' proverbial commitment to enlightened endeavors like democracy and philosophy coexisted with traditional, "irrational," or "superstitious" beliefs in the reality of gods and other supernatural or magical forces.[33]

Again, it is essential to note the disproportionately large contributions that were made by elite families to experience in this realm. For example, through liturgies and other outlays, wealthy Athenians commonly sponsored sacred buildings, choruses, offerings, and other ritual activities, both local and national. In all likelihood, signature events at Athena's Great Panathenaia, like the contests for chariot-dismounters (*apobatai*), javelin-throwers-on-horseback (*hippakontistai*), and armed dancers (*purrhikhistai*), were funded and contested exclusively by members of elite households. And even under the democracy, members of particular aristocratic families continued to provide the hereditary priests, priestesses, and other ritual officials for some of the most prominent cults in the *polis*, including those of several eponymous heroes, Demeter at Eleusis, and Athena Polias herself.[34]

That said, these elite priests and priestesses were not part of any professionalized clerical hierarchy. There was no such body of credentialed specialists in classical Athens.[35] Indeed, in so far as religious behavior was regulated in the *polis* at all, perceived violations were considered along with more secular offenses by regular panels of civilian jurors in the democratically constituted courts of Athens. And in theory at least, any citizen could prosecute any other for crimes like *asebeia* ("impiety") which might have somehow harmed relations between the community and its gods, as we see all too clearly in the notorious trial of Socrates.[36]

The Turn to Culture

To conclude this overview of the consensus account of the Athenian way of life, one should note a palpable shift in practice, perspective, and interest that has taken place in the field of ancient Greek history in recent times. While our predecessors in the field were concerned almost exclusively with the recovery and analysis of observable materialities, like events, practices, and institutions, there has been a pronounced turn towards the study of cultural phenomena over the last thirty or so years. Like our counterparts elsewhere in the discipline, we are now less interested in simply documenting and analyzing the material contents of the political, social, and economic realms. Taking our leads from the works of cultural anthropologists and critical theorists, we are altogether more committed to recovering the contents of the human minds that apparently designed and/ or acted upon those materialities, exploring the ways that the objectively real conditions of existence were construed and represented in prevailing values and beliefs, mentalities and rationalities, ideologies and discourses.[37]

And this cultural turn in our practice has been particularly marked in Athenian studies, where publications now proliferate on, say, conceptualizations of self and other, on the invention of traditions and the formation of social memories, on dominant constructions of genders and ethnicities, on representations of bodies and sexualities, on the objectification of the citizen and non-citizen, the public and the private, the human and the divine, and so forth. And all of these ventures into cultural history have collectively raised the possibility of a new kind of account of Athenian experience, an "emic" or "insider" account, one that effectively parallels and complements more traditional "etic"/"outsider" studies of the material contents of the political, social, and economic realms.[38]

"Democratic Athens"

But whether we today produce histories of material phenomena, histories of cultural phenomena, or some combination of the two, we all still take it for granted that these phenomena can all be mapped onto the same basic, universal model or template of social being. Whatever the differences of opinion that may divide us, we all tend to assume that the *polis* of Athens is most productively conceptualized in abstract, spatial terms as a disenchanted, functionally divisible societal terrain, one that was inhabited by natural psycho-physical individuals. Within this terrain, we tend to reduce what the Athenians called their *politeia*, their entire "way of life," to a more or less discrete, specialist "political" field or system. We then tend to assume that this *demokratia* was a kind of distant ancestor of own liberal governments of today, seeing it invariably as a "democracy," and analyzing it as if it were an essentially egalitarian exercise in power-sharing between free, rights-bearing male "citizens." And we generally take it for granted that this political system exerted a preponderant influence over the social, the economic, the religious, and all other fields of experience, both sacred and secular, thereby defining the distinctive ethos or character of the *polis* as a whole. Thus, the art and architecture of the Athenians, their street patterns, their burial practices, their rituals and their spectacles, their music and their drama, their navy, their hoplite phalanx, even their constructions of the body, courage, and time itself, can all appear to us to be in some essential, irreducible sense "democratic."[39]

The net outcome of all these shared presuppositions, then, is a consensus general vision of the Athenian way of life, one that informs the standard modern literature on the subject. For convenience, we can call this account simply "democratic Athens."

But for all its axiomatic status, this modern translation of ancient experience, this shared common sense vision of a "democratic Athens," is quite deeply problematic. The problems in question are of two distinct kinds. The first are issues that are more readily apparent, namely the problems that arise when one takes this consensus account on its own modern, social scientific terms. The second are more fundamental, more theoretically complex issues, problems that reside in the modern terms themselves. As we shall see in the following two chapters, major problems of both kinds have yet to be adequately confronted, never mind resolved.

2

A World of Contradictions

EVEN IF ONE accepts in principle the use of the universal template of social being which underpins our "democratic Athens," it is hard to avoid being somewhat troubled by the analytical results. Even if one is prepared to embrace the model and all of its standard social scientific categories, the general account of Athens that it predisposes us to produce contains some rather startling improbabilities. Simply stated, if the Athenian way of life really was defined in the end by an ensemble of progressive, democratically egalitarian, proto-liberal political arrangements, it was a lifeworld that was riddled with a number of obvious and fundamental tensions, paradoxes, and contradictions.

Enlightened Slaveholders

First, and perhaps most obvious of all, if the Athenians were such idealistic democrats, how could they possibly have tolerated the flagrant exploitation of tens of thousands of slaves in Attica? Even if it is hard to determine the exact number of these slaves at any given time, and even if a good number of Athenian households probably owned none at all, few would deny that the Athenian *polis* meets the criteria for a "slave society."[1] Since Athenians did not as a rule purchase one another's labor in the capitalist fashion, it is widely acknowledged that the ranks of the exploited in classical Attica were filled almost exclusively by unfree, mostly non-Greek chattels. So how can this "slave society" have been "democratic" in any meaningful sense?

A number of specialists have attempted to confront this issue directly and explore various possible articulations between democracy and slavery in Athens.[2] But most, it seems, would prefer to minimize, sidestep, or just ignore this apparent contradiction. In so far as they address it at all, they consider only how slavery might or might not have impacted the daily practice of democracy in Athenian institutions. More often than not, the issue can then be reduced to a single, simple question: Would the Athenians have had the leisure and the opportunity to participate in political life in the required numbers without the use of slave labor? Answers to this latter question then inevitably turn on an even narrower issue: How widespread was slaveholding among "peasant" smallholders in the Attic countryside? And since the consensus answer to this last question seems to be "not very widespread," most observers appear content to let the whole matter rest there. While "democracy" and slavery may have coexisted within the same social formation, the former was quite possibly innocent of any meaningful linkage with the latter.[3]

Perhaps this rather evasive conclusion is the result of a certain discomfort with a troubling topic. Whatever the case, it seems to be an inadequately rigorous response to a complex issue. Of course, if one is committed to thinking of the *polis* as a functionally compartmentalized entity, a polity where a democratic, egalitarian field of the political could to a point subsist as an autonomous realm of thought and practice, effectively insulated from and unconditioned by the undemocratic, unegalitarian practices that were going on in the economic field, it might then be possible to see no direct contradiction between democracy and slavery in Attica. And it might then be possible to analytically marginalize the sometimes unseemly activities that went on in the workplaces of "democratic Athens," treating them as if they were really going on somewhere else.

But even then, one must still confront some troubling questions. For example, given that the surplus resources of a slave-owning elite minority were manifestly important to the reproduction of Athenian social being, can one so easily disentangle slavery from the overall way of life of the *polis*? Moreover, if the practice of *demokratia* in Athens was not "theoretically" inconsistent with the routine exploitation of slave labor, as it evidently was not, then what exactly was the theoretical basis of this *politeia*? And if the theoretical premises of Athenian government thus did not resemble the post-Enlightenment egalitarian ideals which undergird modern liberal government, as they evidently did not, why do we insist on treating them as two variant forms of the same type of political regime?

Democratic Imperialists

Similar questions and issues are prompted by the second of our contradictions, namely that between Athenian democracy and Athenian empire. If the modern consensus would read Athenian *demokratia* as a proto-liberal exercise in "direct democracy," as a regime committed to the political equality and autonomy of individuals, how then would historians reconcile that commitment with an imperialist domination of as many as 170 other Greek *poleis* around the Aegean basin during the latter half of the fifth century?

Here, scholarly discomfort is perhaps more explicitly apparent. Indeed, a certain anxiety about Athenian imperialism has prompted a fair degree of special pleading over the years. Perhaps, some would claim, the harshness and the exploitative character of the Athenian empire has been overstated. Perhaps it really was not an "empire" in the strictest sense of the term, but a less coercive form of "hegemony." Besides, it is not as if Athenian sway had no positive outcomes for others. Indeed, we are sometimes reassured, there are signs that it was actually quite "popular," that it furnished benefits, material and/or political, for at least some of its subjects, notably those upon whom the Athenians imposed "democracies," however circumscribed. Either way, one could argue, domination by the Athenians was at least preferable to the most likely alternative, namely domination by the Persians.

Then again, if all this fails to convince, it is always possible to blame some of the less palatable features of the empire, or the imperialist project as a whole, on the undue influence of certain "demagogues" or certain sections of the citizen body, like "the poor," who were apparently all too happy to better themselves at others' expense. One can always believe that sectional interests of one kind or another caused the Athenians to behave in ways that were somehow untrue to themselves. Or more positively, one can always stress that the cause of democracy in Athens itself was directly advanced by empire, not least by the significantly higher public revenues that were available to fund the stipends of jurors and *polis* officials.[4]

Has any other ancient empire, one wonders, stirred quite this level of apologetic fluster? This in itself is revealing. As for the arguments themselves, there may be a qualified truth in each of them. But even if every one were completely true, the perceived inconsistency or contradiction between the supposedly democratic principles and an imperialist project would not be conclusively resolved. It would merely be justified or otherwise explained away.

So why, one must ask, is all this analytical work focused only on one side of the contradiction and not on the other? Why does a case like this not cause us to question how we construct "Athenian democracy" in the first place? Why, again, do we never seem to entertain the possibility that *demokratia*—the very same regime that the Athenians saw fit to impose on a number of their not so free, not so equal subjects—might have been categorically different from our own political arrangements? Why do we never consider the possibility that its ethical foundations, whatever their nature, might not in fact have been so at odds with the exigencies of empire after all? Again, it appears, we would prefer to imagine that the Athenians inhabited a modern-style, "rationally" compartmentalized lifeworld, one where multiple discrete fields of experience were somehow able to coexist, despite being sustained by mutually contradictory logics and principles. To maintain our faith in the possibility of an egalitarian, proto-liberal political realm in Athens, we must again assume that it was somehow isolated and insulated from other manifestly less egalitarian realms of experience. In other words, we must assume that the commitment to democracy by the otherwise enlightened Athenians was, for some as yet unexplained reason, highly selective.

Women Excluded and Secluded

And much the same could be said of our third contradiction, which seems to reveal a highly selective commitment to democracy even within the *polis* itself. For if the Athenians ordered their world according to something like post-Enlightenment, liberal principles, how could they have excluded their own womenfolk, the native-born, adult females of Attica, from political institutions and denied them their civic entitlement to self-determination? Can we even speak meaningfully of female "citizens" in Athens at all? Apparently, the commitment to democracy in Athens was confined exclusively to the public realm of government, having no purchase or meaning within the private spaces of households, the sources of all Athenian livelihoods.

Far from evading or denying this seemingly egregious contradiction in Athenian social being, specialists have long been deeply troubled by it, especially since the 1970s, when a commitment to the cause of women's history first began to make a deep impression on the field. Indeed, one particularly influential current of thought has expressly sought to link the rise of democracy in *poleis* like Athens to the advent of an ever more repressive "patriarchy" or "reign of the phallus." According to this argument, the "misogynistic" males of classical Greece willfully and systematically stripped females of the higher "status" or "value" that they had allegedly enjoyed in archaic times, denying them their "full rights" as "citizens," banishing them from public life, and "secluding" them away in the "domestic sphere," where they were free to do little more than be seen, exploited, and dominated. Only in anomalous, "conservative" Sparta, it seems, did women retain some vestige of their earlier social power and standing.[5]

In recent years, however, historians schooled in the more theoretically demanding ways of gender history have significantly complicated this picture, producing a more nuanced, more compelling alternative to the received wisdom. Though it might draw on much the same body of evidence, this alternative account would read the ancient testimony somewhat differently, trying to understand the Athenian way of life more on its own pre-modern terms. The result is an emerging, composite picture of female experience which foregrounds the active and vital roles played by women in areas like household management, child-rearing, neighborhood policing, property disposition, citizenship determination, manufacturing, banking, commerce, and religion. Indeed, it now seems that warfare and public decision-making were the only essential, life-sustaining activities in which women were not active participants.[6]

Besides, however limited or insignificant female contributions to the life of Athens might seem to our modern eyes, it is clear enough that the Athenians themselves saw women as full members of the *polis* community in their own right. As supporting evidence for this claim, one need only note the currency of the term *politis* (plural *politides*), the feminine counterpart of the masculine form *polites* (plural *politai*), the word we conventionally translate as "citizen." In other words, the Athenian *polis* was not an enterprise that existed largely to serve the needs of its freeborn, adult, male constituents. Females in Athens may not have been interchangeable with males. But as *politides*, they were by definition members of the *polis*, even if the contents of their membership were obviously somewhat different. As the Athenians themselves would put it, they too, like their fathers and husbands, "had a share in (*metekhein*) the *politeia*." And like their fathers and husbands, they too could suffer the loss of part or all of that share in the *politeia* and be consigned to a state of "honorlessness" (*atimia*) if they seriously threatened the well-being of the *polis* by, for example, submitting to the advances of an Athenian "seducer" (*moikhos*), thereby setting the bloodline and integrity of their *oikoi* at risk. But where male *atimia* might involve, say, the forfeiture of one's entitlement to address the assembly or enter the *agora*, the female equivalent might require divorce, expulsion from the marital *oikos*, and a ban from entering the sanctuaries of the gods.[7]

Suffice it to say for now, when we consider female experience in Athens, we seem to be posing the wrong questions, finding problems and contradictions where they may not exist. The real question at issue here is not: Why did the democratic Athenians deny their womenfolk their full political rights as citizens? Rather we should be asking: If women could be considered full members of the *polis*, how exactly was "membership" conceptualized at the time, and how did it differ from our modern, narrowly legalistic notion of "citizenship"? To put it another way, if "having a share of the *politeia*" did not necessarily involve voting in the assembly, but obviously could involve, say, managing an Athenian *oikos* and honoring gods of the *polis*, why do we historians insist on defining this *politeia*, this *demokratia*, as an ensemble of specifically political arrangements, as a "form of government" that was functionally equivalent to our own modern, secular "democracy"?

The "Paradox of the Polis"

Finally, supporters of the mainstream "democratic Athens" account must at some point confront what would seem to be a significant contradiction within the very heart of this putative ancient "democracy." I refer to the conspicuously unequal distribution of resources among male "citizens," especially the significant wealth gap which separated the super-rich minority from most of their fellow Athenians. How then can one reconcile this socio-economic inequality with the alleged "egalitarianism" of the *polis*, with the notion that all Athenians were somehow "political equals"?

One distinguished observer has called this apparent contradiction "the paradox of the *polis*."[8] And indeed, far from avoiding the issue, scholars tend to take it for granted that wealth asymmetries among male citizens were a problem, a potentially serious source of tension or instability within "democratic Athens" and elsewhere in the Greek world.[9]

Among works which have attempted to address this problem directly, the most influential have probably been a series of studies by Josiah Ober. For Ober, the ultimate source of long-term socio-political "stability" in democratic Athens was the "domination by the mass of ordinary citizens of the conceptual apparatus essential to decision-making." True, the de facto "leaders" who guided assembly deliberations were invariably wealthy elites. But to become accepted as a leader in the first place, Ober contends, one had to use rhetorical tropes and devices which conformed to the prevailing "egalitarian," "democratic," or "mass" ideology, thereby ensuring the production of outcomes that favored the interests of non-elite Athenians. This "ideological hegemony of the masses" thus "counterbalanced" the "social power" of the affluent minority, preventing the formation of an elite oligarchy in "the political realm." As a result, Athenian government was possibly the closest that "we" have come to a "real democracy," where a "relatively broad citizenship of ordinary (non-elite) men" assumed "true political authority" and "directly ruled a complex society."[10]

As appealing to modern sensibilities as this vision of "real democracy" might seem at first sight, it again begs as many questions as it answers.

Most obviously, if the sectional self-interest of the poorer "masses" exerted a preponderant influence on Athenian decision-making in venues like the assembly and the council, why is it that the visibly uneven distribution of wealth among Athenian households remained essentially unchanged throughout the classical era? Moreover, in a world where elites all but monopolized channels of public speech and positions of leadership, how exactly were non-elites capable of formulating their own independent sectional interests, never mind their own distinctive "ideology," in the first place? Where exactly is the evidence that the poorer majority was actually capable of self-organizing or even self-identifying as an autonomous, united interest group, never mind as one capable of imposing its own common sense vision of the order of things upon the thought and speech of all Athenians?

Besides, there are numerous indications elsewhere in the fabrics of *polis* experience that the supposedly egalitarian Athenians were a good deal less troubled by manifest inequalities between rich and poor than we historians tend to be today.

As far as we can tell, they had little or no problem with a distribution of military labor which would always mirror the distribution of wealth among households in Attica. They had no problem with the idea that the elite minority would provide all the generals, all the cavalrymen, and the best-equipped hoplites, while poorer Athenians would perform altogether less glamorous, less prestigious military roles, usually serving as light-armed troops and as paid oarsmen in the navy's triremes, often alongside slaves. Likewise, they seem to have been entirely comfortable with the idea that elites would assume responsibility for prosecuting all *graphe* cases, where the *polis* as a whole was deemed to be the injured party. Presumably too they just accepted that they lived in an environment where elite families would retain control of major priesthoods, sponsor spectacles at major festivals, fund sanctuary monuments, and otherwise perform a conspicuously prominent role in the ritual life of the community. And if they actually did have a problem with these and all other such public expressions of inequality, why did the allegedly regnant poor majority do nothing to change, neutralize, conceal, or eliminate them?

A Sociological Improbability

Pulling all this together, our standard account of "democratic Athens," even when taken on its own terms, appears to be seriously problematic. Simply stated, if life in the Greek world's largest, most influential *polis* was ultimately shaped and conditioned by the presence of an egalitarian, proto-liberal "democracy," then this life was continually confounded by all manner of paradoxes, tensions, and dissonances that cannot easily be rationalized or explained away.

Of course, to a point, one can agree with Edward Cohen that "complex and dynamic civilizations" will always produce contradictions, anomalies, and exceptions to the rule.[11] But at what point, one wonders, does the sheer number and significance of the "undemocratic" features of "democratic Athens" start to threaten our very belief in the existence of any such "rule"? Maybe one really can make a case that practice in the assembly itself was governed by some kind of democratically egalitarian ethos or ideology, even if elites monopolized all leadership roles. But the prevalence of any such ideology elsewhere in *polis* experience is far from self-evident. On the contrary, one could fairly make the case that inequalities of various kinds were not just visibly represented but actively reproduced in family structures, in workplaces, in military practices, in religious spectacles and monuments, and in relations with numerous other Greek *poleis*. In other words, when one thinks of all the mechanisms which secured the very lives and livelihoods of the Athenian people, few if any can be said to be distinctly and unequivocally egalitarian. When one considers the totality of the Athenian way of life, "democratic" practices would seem to be much more the exception than the rule.

To all of this, one might still respond that our own enlightened, modern liberal polities are hardly innocent of such egregious contradictions themselves, as if that would make the Athenian case somehow less remarkable or more excusable. To be sure, the professed commitment of modern nations like Britain and the United States to securing the rights, freedoms, and equalities of humans as individuals has historically coexisted with aggressively imperialist policies, the exploitation of slave labor, the exclusion of women from the privileges of full citizenship, and a tolerance for extreme economic inequalities among citizens. But there is at least one fundamental difference between the ancient and the modern instances.

In western modernity, influential voices have almost always been there to draw critical attention to such contradictions and ensure that they are not conveniently ignored. And not infrequently, such critique has led to direct, concerted action. To take only the American example, one can hardly overstate the profound historical consequences of projects and initiatives like the abolitionist movement, the women's movement, workers' strikes, civil rights campaigns, and anti-imperialist protests. Indeed, one could say that the story of the modern United States has been substantially shaped and defined by these attempts to resolve the more flagrant contraventions of the ideals upon which the polity itself was founded.

The ancient Athenian case could not be more different. There never were any organized campaigns, protest movements, or other concerted efforts that aimed to resolve the manifest contradictions of our "democratic Athens." More revealing still, the Athenians do not even seem to have been aware that any such contradictions were actually there in their *polis* at all. No doubt, one can find evidence that certain ancients were in some way troubled by, say, poverty, slavery, or Athenian imperialism. But crucially, there is no evidence that such "problems" were deemed problematic specifically because they contradicted the basic ethos or principles of the Athenian *politeia*. To my knowledge, no one ever claimed that the Athenians' treatment of their womenfolk, their slaves, their imperial subjects, or their poor was wrong because it was essentially "undemocratic."

In sum, there is something sociologically improbable about "democratic Athens," the general account of the Athenian way of life that one finds in almost all modern text books and scholarship. When it historicizes *demokratia* as "democracy," this account effectively requires the Athenians to inhabit an extraordinarily convoluted, almost self-defeating lifeworld, one where political experience was demonstrably and continually at odds with experience in all other societal fields. When one thinks about all the contradictions that this account produces, one has to wonder not just whether a democratic form of government really was the defining essence of Athenian social being. One has to wonder whether anything resembling our own modern liberal democracy was really there at all.

Such are the kinds of problems that arise when we critically scrutinize the "democratic Athens" account on its own terms. When we then turn to question the terms themselves, when we critically scrutinize the presuppositions and categories of the template upon which our "democratic Athens" is premised, we find problems of an even more serious nature and magnitude, as we shall now see.

3

Missing Objects

TO SPEAK MEANINGFULLY of a "democratic Athens," a world ordered by and around an ancestor of modern liberal democracy, presupposes the real presence in Athenian lived experience of a complex ensemble of rather specific phenomenal conditions. Most immediately, it presupposes the presence of a community of human subjects who understand and objectify themselves as an aggregate of naturally autonomous individuals, each one endowed by nature with the kinds of personal rights, freedoms, and equalities that a democracy by definition exists to nurture, express, and protect. Accordingly, it presupposes the presence in Athens of phenomena like citizenship, private property, and bounded realms of state, society, economy, and religion, phenomena which expressly enable and encourage subjects to flourish as self-actualizing individuals. And it thus presupposes the realization in Athens of a common sense conception of social being as a largely secular societal terrain or arena, one that is functionally divisible into discrete fields of experience. In short, to speak meaningfully of a "democratic Athens," one would have to show that lived reality in this ancient *polis* corresponded quite precisely to the prescriptions of the standard modern analytical template discussed earlier.

The problem is that we have little or no explicit evidence for any such correspondence between ancient lived reality and modern social theory. There is no incontrovertible evidence that any of the various phenomena listed above, the essential conditions of any democracy's possibility, were really there as such in Athenian experience. As far as we can tell, the Athenians knew nothing of the social objects we call state, society, economy, and religion. They had no terms that were exactly equivalent to our modern, western notions of citizenship or rights. They had no words that precisely capture our ideas of privacy or secularity. And there is no evidence whatsoever that they objectified humans in general as autonomous, self-actualizing beings, all born with a natural entitlement to enjoy freedom and equality as individuals.

Of course, one might claim in response that most if not all of these phenomena are all either already there or potentially there in every historical environment. After all, this much is surely taken for granted by our standard analytical template itself, which continues to inform and shape the production of most if not all knowledge throughout the discipline of history as a whole. Thus, our textbook accounts of complex past worlds almost always begin precisely by distinguishing between, say, political and social fields, between public and private realms, and between sacred and secular experiences. Our master narratives tend to be explicitly structured around processes like state formation, economic development, and secularization, regardless of whether these processes were known or visible as such to our subjects at the time. And surely the discipline's conventional, near-universal use of this general model and its master categories would be fully endorsed by the ultimate authority on such matters, namely our mainstream social sciences, which likewise take it for granted that, say, political science, sociology, and economics will always have their own discrete, materially self-evident objects of study.

No doubt, this is all true. But it is also true that the fabric of lived experience in classical Athens does not readily yield to our standard analytical devices. Indeed, I would argue, the only reason why all of the above phenomena might seem to us to be self-evidently there, as if already imprinted in that fabric, is because the basic tools of our practice predispose us to see them there. If we were somehow just to study the Athenian *politeia* on its own terms, without our standard inventory of categories shaping our expectations, we would find no actual evidence as such for any of these objects. They would all be missing. For if the evidence can be said to confirm anything either way, it is that these various phenomena would have made little or no sense as things-in-themselves in classical Athens.

A Society of Citizens?

To illustrate the point, let us first consider one of the more primordial divides that we conventionally inscribe on the Athenian terrain, namely that which separates state from society, the political from the social, and thus to a point the public from the private.[1] Again, it is not just the case that the Athenians had no terms that corresponded exactly to these phenomena. It is also the case that one cannot detect their presence even as unseen forces, visible only through their effects on the evidence of Athenian experience. And this may not be too surprising, since a state-society or public-private divide would have been all but unintelligible in this particular historical context.

To begin with, as we saw in Chapter One, constituents of Athenian "society" were necessarily present everywhere in what we might presume to call the

Athenian "government" or "state." In Athens, there were no professionalized governmental agencies or ministries, no specialist administrators, no bureaucratic experts, no permanent public authorities of any kind. In the assembly and council, the leaders, speakers, and voters were all what we would call civilian volunteers. In the courts, all cases were overseen, initiated, prosecuted, and judged by "laymen." Nor was there any formal apparatus of police. Essentially, it was left to the people of Attica, male and female, Athenian and non-Athenian, to regulate themselves on a day-to-day basis. Likewise, there was no professionalized standing army or navy. All commanders, cavalrymen, infantrymen, and oarsmen were again, by modern standards, untrained amateurs. So too, when it came to maintaining the all-important ritual engagements with the pantheon of Athenian gods, with the beings upon whose continuing favor the lives of all humans in Attica depended, the task was left to non-specialists. For whatever reason, the Athenians never saw fit to formalize a clerical hierarchy or cohort of ritual professionals to manage human-divine relations on behalf of the *polis* as a whole.

In sum, there was no modern-style sense in Athens that life in general was being continually managed and administered by a free-standing, superordinate apparatus of "government." Thought and practice did not conspire to conjure up the effect of an automaton-like "state" that was somehow separate from an Athenian "society."[2]

When we then consider the nature and sources of the many resources which were required to administer and sustain the life of the *polis*, the possibility of any meaningful state-society divide in Athens recedes yet further from view. The Athenians were quite capable of distinguishing between *idia*, things that belonged to a particular person or *oikos*, and the *koina* ("things held in common"), *demosia* ("things of the people"), or *politika* ("things relating to the *polis*") that belonged or pertained to the community as a whole. But this general distinction could not and did not correspond to any abstract, spatial division in the societal terrain between discrete public and private realms or political and social fields.

Thus, it was simply taken for granted that *idia* resources could and would be mobilized, through mechanisms like liturgies, to serve the common good of the *polis*. Likewise, as we saw above, the deliberative process in the assembly presupposed the presence of individuals who had the leisure, knowledge, rhetorical ability, and other personal resources necessary to provide *demos* with regular, informed advice. So too in the law courts, *graphe* procedures against Athenians who harmed the *polis* presupposed the availability of individuals who had the time, energy, and skills necessary to prosecute the cases, along with the wealth necessary to cover the large fine if the suit were to fail. And in much the same way, deme registrations, the all-important processes for enrolling eighteen-year-old males as members of *demos*, presupposed a network of relatively stable, integrated neighborhoods and localities, where demesmen could readily acquire the information necessary to determine whether or not one was the product of a legitimate Athenian *oikos*.

More generally, the management of everyday behavior in Attica presupposed an entire community of subjects who possessed the will, interest, knowledge, and interpersonal connections necessary to police themselves and, if necessary, initiate court proceedings, gather evidence, summon witnesses, and so forth. So too, the conduct of warfare in Athens presupposed a body of *oikoi* who could between them furnish all the necessary tactical and strategic know-how, weapons, body armor, horses, ships, oars, and other materiel necessary to mount successful military operations. And the ritual life of the *polis* presupposed a population at large which was prepared to do all that was appropriate and necessary to optimize relations with dozens of different divinities, a population which was ready to maintain a more or less constant traffic in offerings of all kinds in their houses and at shrines, sanctuaries, and any number of other communal sites.

If "society" was thus continuously and necessarily present in any putative space of an Athenian "state," there was also little or nothing to stop this "state" intruding into whatever we might presume to call an Athenian "society." In Athens, there was no formally codified bill or declaration of rights, a *cordon sanitaire* which might have secured a free realm of privacy from any such intrusion, not least because such a mechanism would again have made little or no sense at the time. The essential logic of modern liberalism, whereby the powers of an autonomous "government" must be expressly circumscribed and contained so as to maximize the freedom of the natural individual to prosper, would have been entirely alien to the Athenians. This was, after all, a world where the actions of *demos*, the ruling agency, were determined and authorized entirely by civilian volunteers, including one's husband, one's father, or oneself. So even if one could somehow imagine a free-standing "government" along these lines, as an agency somehow separate from one's "society," why would one even need to claim "rights" of privacy that would protect one against any such entity?

Moreover, the Athenians had no equivalent of our notion of absolute property ownership, the very idea upon which any free realm of privacy would presumably be premised. As Alison Burford has put it, the *polis* community remained the "proprietor-in-chief" of all lands within its compass. Strictly speaking, there was no equivalent in Greek of our term "landowner." There was no conception of "property" as such, as an abstract, readily commodifiable, indefinitely exchangeable resource. The *idia* holdings of each Greek *oikos* were typically conceptualized as "parts" or "shares" of a whole (*moirai, moria*), as "allotments" (*kleroi*) from a communal body of land that had at some point been divided and distributed to sustain the vitality of every family lineage that constituted the *polis*. Accordingly, far from insulating a free realm of privacy from governmental interventions, the Athenians saw securing the "private" well-being of each *oikos* as one of their gravest "public" responsibilities.[3]

For example, it was the express, sworn duty of the chief or "eponymous" archon in Athens to oversee matters of inheritance and ensure the survival of every Athenian household, to guarantee that the holdings of each one would be the same at the end of their year-long terms of office as they had been at the beginning.[4] More striking still, one could be indicted in Athens by any fellow Athenian for harming the *polis* as a whole if one were to mismanage one's own *oikos* or somehow threaten its continuing well-being. Thus, the list of attested "public interest" procedures (*graphai*) that could be initiated by any male member of the *polis* includes suits against those guilty of selling their bodies as prostitutes (*hetairesis*), seducing female members of other *oikoi* (*moikheia*), mistreating their parents and other family members (*kakosis*), and dissipating their inheritances through laziness (*argia*) or insanity (*paranoia*). In such an environment, I submit, where there seems to have been no self-evident distinction between the well-being of the *polis* and the well-being of its constituent *oikoi*, the application of modern categories like state and society, political and social, and public and private will not just be problematic. It will be all but meaningless.

As noted in the first chapter, specialists are not unaware of these "category problems." In recent times, a growing number of historians have come to recognize that Athenian experience to a point defies and confounds our standard analytical terminology. As a result, there is now an emerging consensus that we should think of "the Greek *polis*" in general as some sort of hybrid entity, as something that was somehow simultaneously a state and a society, a kind of inseparable blurring or fusion of the two, a self-regulating "political society" or "society of citizens."[5]

But these alternative formulations present their own difficulties. For a start, there is a rather immediate philosophical problem here. How can one logically or meaningfully speak of the *polis* as a "fusion" of two phenomena, a state and a society, that had no attested independent existence as entities in the first place? And if there was no state or society there to begin with, what was a *polis* "made of" exactly? Moreover, even if we overlook for a moment their dependence upon conventional, modern terminology, formulations like "political society" and "society of citizens" seem only to complicate matters still further. They simply gloss over the fact that the vast majority of the inhabitants of this putative *polis* "society" were not actually "citizens" in our sense, since they took no part in "political" life at all. And they likewise fail to acknowledge the fact that even those who were "citizens" were not free to act individually as "political" agents as and whenever they wished. For the most part, Athenians could produce authoritative, universally binding decisions only under very specific circumstances, like assembly meetings, when gathered together as *demos*, "the people" of the *polis*. In other words, our conventional categories do not allow us to capture this essential distinction. They are incapable of representing a world where the same person could function in effect as a particular member of a particular household at one moment and as a human incarnation of the *polis* itself the next.

A Liberty of the Ancients or the Moderns?

The absence of a state-society divide in Athenian experience in turn creates problems for our two most prevalent accounts of freedom (*eleutheria*) in Athens, the freedom to "live as one likes."

The first of these corresponds roughly to what Benjamin Constant in 1819 called the "liberty of the ancients," defined as a "sharing of social power among the citizens of the same fatherland," such that each one was "almost always sovereign in public affairs, but a slave in all his private relations."[6] According to this account, in other words, Athenian *eleutheria* was essentially a "collective freedom," whereby "the authority of the social body" continually "interposed itself and obstructed the will of individuals."[7] As a result, in Athens "[t]here were no theoretical limits to the power of the state, no activity, no sphere of human behaviour, in which the state could not legitimately intervene provided the decision was properly taken for any reason that was held to be valid by a legitimate authority."[8]

However, Constant himself was convinced that this "enslavement of individual existence" was a good deal less "complete" in Athens than it was in other ancient polities.[9] Even if the citizen of Athens was still "much more subservient to the social body . . . than he is in many of the free states of Europe today," he still enjoyed a measure of "individual independence" and a certain "security in private pleasures" that are the hallmarks of liberal freedom, the "liberty of the moderns."[10] And others have come to similar conclusions in the years since, most notably Mogens Hansen.[11]

In short, these collectivist and individualist accounts of Athenian liberty both require us to think of freedom in spatial terms, as the property of a particular field or sphere of experience. Both presuppose the existence of a functionally differentiated societal terrain. Just as the collectivist account assumes the presence of a free-standing state agency that can act in and upon something that can be called society, so the individualist account takes it for granted that we can determine precisely where a public sphere of the political ended in Athens and where a free private sphere began. Yet as we have seen, there is no direct evidence from Athens to support either of these assumptions. Which is to say, our accounts leave us unable to explain the essential nature of the Athenian freedom to "live as one likes," a freedom that apparently did not depend upon the inscription of modern-style functional divisions in any abstract societal terrain.

An Athenian Economy?

Very similar problems arise when we then go in search of evidence for an Athenian "economy." Here again, conventional modern wisdom offers us a pair of contrasting alternatives. Advocates of a "formalist" account insist that one can in fact observe the quasi-mechanical workings of an economy in Athens, a more or less discrete "system" of production, distribution, and exchange that was governed by something like "market forces." As in our capitalist modernity, the claim goes, wealth creation in the classical era was increasingly pursued for its own sake by a population of self-interested, utility-maximizing individuals in its own discrete field of experience, one from which government was effectively excluded.[12] By contrast, "substantivists" would contend that the processes of production, distribution, and exchange in classical Athens were still largely conditioned by non-economic concerns, like honor, status, and citizenship. So far from being a free-standing, self-ordering system, the Athenian economy was still "embedded" in other fields of experience, like the political and the social.[13] But again, both of these accounts are problematic in their different ways.

The formalist account seems to underestimate the specificity of the conditions that are necessary to produce and sustain the existence of a complex social object like an economy. To be sure, there were activities in Athens, like banking, insurance, and commercial commodity production, which look like relatively straightforward exercises in wealth accumulation. Doubtless, ancient bankers, insurers, and manufacturers were just as interested in maximizing their incomes as their modern capitalist counterparts. And doubtless, the prices they charged for their various goods and services were to a point determined by considerations of what we might call supply and demand. But the mere incidence of such phenomena does not in itself oblige us to conclude that life in Athens was therefore governed by the workings of a dedicated "economic" system, a mechanism exclusively concerned with the allocation of what we moderns deem to be "valuable" material resources.

To begin with, it is impossible to draw any meaningful distinction between an economic and a social field in classical Athens. In Greek antiquity, there were no specialist commercial entities, no limited liability corporations or other bodies that were formed exclusively for the purpose of accumulating wealth. Communally administered revenues of the *polis* aside, all wealth in Athens was produced and accumulated by households, which were simultaneously what we would call economic and social units, the sources of all livelihoods. In such an environment, the very idea that one might somehow create a legally fictitious household or association expressly to generate an income stream would have been anathema if it were even conceivable. Moreover, in a world where households were commonly supported by ancestral plots of land, passed on from one generation to the next, an economy based on a freewheeling, capitalist-style market for commodifiable "property" simply could never have materialized. Likewise, in an environment where the native inhabitants conventionally worked only for their own households, not for those of others, there could be no true capitalist-style market in "free labor." And as a result, there could be nothing resembling the modern sense of economic interdependence that arises when citizens routinely sell their labor to one fellow citizen in order to acquire the means to buy goods from others. Each *oikos* in Athens was its own miniature economy, so to speak, much as the Greek term *oikonomia* ("household management") originally indicated.

Then again, the possibility that an economy might have materialized in Athens as a discrete, free-standing system or realm of experience was further deterred by the complete inseparability of wealth-producing *oikoi* from the *polis* to which they all collectively belonged. So long as maintaining the well-being of each individual *oikos* remained an express concern of the *polis* as a whole, there could be no a priori divide there between a state and a society, a political and a social realm, or public and private spheres. And thus, as we saw earlier, when the *demos* acted to safeguard the livelihoods of Athenian households, it did not do so in the indirect capitalist fashion, by, say, lowering taxes or interest rates, promoting competition among *oikoi*, or otherwise manipulating the general conditions for wealth accumulation. It did so by acting in ways that more directly promoted the vitality, integrity, and continuity of each particular household as a self-supporting unit. In other words, one cannot meaningfully separate an "economic" realm from a "political" realm in Athens.

More generally still, the Athenians knew nothing of the various modern discursive phenomena that help to constitute "the economy" as a self-evident, measurable object of analysis in the first place. Theirs was an environment entirely unacquainted with reified abstractions like "labor" and "capital," "property" and "the market," "supply" and "demand," or "economic sectors" like "manufacturing," "trade," and "agriculture." They knew nothing of complex metrological devices like a "standard of living," a "balance of payments," or a "gross domestic product." Indeed, unlike ourselves, they were not in the habit of using any kind of statistical aggregates or averages to measure, monitor, and manage the material fortunes of any abstract "population" as a whole. And this is in the end because it simply would never have occurred to them that the overall vitality of their world could actually be determined by some secular, materialist, narrowly economistic standard of "value."

When they convened to deliberate on the general welfare of the *polis* in settings like the assembly, the Athenians were of course quite capable of discussing basic material concerns, like grain supplies, imperial revenues, or the common funds available for military campaigns. But there is no indication that they believed these concerns were subject ultimately to an array of largely unobservable, macro-level "economic" forces, conditions, and processes like "output" and "productivity," "inflation" and "deflation," "growth" and "recession." Revealingly, in what is arguably the closest we ever come to a "state of the *polis*" address in Athens, the 431 funeral oration reported by Thucydides, Pericles attributes the general vitality of the Athenians at that moment to the distinctive features of their ancestral "way of life" (*politeia, epitedeusis*) as a whole, such as their obedience to laws, written and unwritten, their mutual tolerance and forbearance, their aversion to physically grueling, Spartan-style methods of education, their fondness for collective deliberation, and their generally selfless commitment to a common good. He barely mentions phenomena that could be measured by a modern, economistic value standard at all.[14]

And all of this is to say nothing of the inextricable entanglement of any putative "economy" with ritual activities in classical Athens. This, after all, was a lifeworld where the community's principal treasuries belonged technically to gods. It was a lifeworld where the community's single largest ongoing expense was its investment in "irrational" gifts of all kinds for divinities, from innumerable daily offerings of foodstuffs, liquids, and animals, to lavish theatrical and athletic spectacles, to temples and other costly monuments. And as Athenian rituals and festivals repeatedly attest, it was a lifeworld where divine agencies ultimately controlled life's most essential concerns, like birth and death, health and wealth, fertility and happiness, failure and success. In short, the idea that there might have been some wholly secular, narrowly materialist system of "value" allocation operating somewhere out there in the fabrics of experience, unobservably determining the well-being and prosperity of all, would have been all but unthinkable to the classical Athenians.

If there are thus serious difficulties with the formalist account of an Athenian economy, the substantivist alternative is no less problematic. But here again the problems are more philosophical or theoretical in nature.

For a start, if "non-economic" fields like the political and the social are no more self-evidently visible in Athenian life than the economic, what exactly would an "embedded" economy have been embedded in? How helpful or productive is it to speak of this economy as being somehow immanent in a political realm and a social realm when those realms were themselves still "fused" together, still awaiting the materialization of the state and the society that would define them as distinct realms in the first place?

More generally, once we concede that the Athenians themselves were oblivious to the workings of this putative economy, and thus did not organize their lives upon the assumption of its existence as an entity, how exactly does our assumption of its existence help us to understand and make sense of those lives, which is surely the object of the whole exercise? Then again, in the absence of any explicit evidence for an economy as such, what makes us so sure that it was really there at all in any sense, aside from our own modern analytical predisposition to see it there? And if this predisposition is continually resisted by the ancient evidence, at what point should we begin to question it? If the Athenians themselves give us no cause to draw a distinction between "economic" and "non-economic" phenomena, so that we in effect have to actively manufacture their economy for them and impose it upon their lifeworld, are we still doing something that can meaningfully be called "history"?

Until advocates of the substantivist position are prepared to confront such questions, to do the hard philosophical work of specifying the quiddity of the economy as an object in ancient experience, we have no compelling reason to suppose that there really was any such thing there at the time, whether embedded, disembedded, or otherwise.[15]

An Embedded Religion?

Much the same kinds of issues also arise when we then try to locate the object we call "religion" in the evidence of ancient experience. As we saw earlier, conventional wisdom takes the presence of this phenomenon entirely for granted in its accounts of the lifeworld of the Athenians. Like state, society, and economy, religion is something that most of us just assume is always already there in more or less any complex historical environment. Yet, as most of us are also willing to recognize, there was no close equivalent to our category in Greek thought or language. Moreover, even if we insist on trying to recover such a phenomenon from Athenian experience, it remains far from self-evident where exactly a sacred realm of religion would end and a secular realm of non-religion would begin.

After all, as we saw earlier, this was a lifeworld where prayers and offerings routinely accompanied almost all human endeavors, including ostensibly "secular" ones like assembly deliberations, court cases, commercial transactions, seafaring, and warfare, presumably because it was felt that superhumans ultimately had some say or interest in the outcomes of all such activities. While there may have been strict rules about how and where one might try to engage directly with divinity, there could be no rules about where in life's many pursuits and processes the gods might choose to intervene. And if there thus could be no truly secular or disenchanted fields in the terrain of Athenian experience, fields from which the all-powerful gods could somehow be excluded, it makes no sense for us to try to delineate the bounds of a sacred realm of "religion" in that same terrain.[16]

As in the case of the Athenian "economy," some scholars are now increasingly willing to recognize that a bounded field or system of religion may not have been so self-evident in Athenian experience after all. But the suggested alternative is to see religion as something else that was simply "embedded" in other fields of experience.[17] And again this proposition only raises further problems.

Most immediately, the mere absence of evidence for a discrete field of religion in Athens does not in itself constitute positive evidence for its "embeddedness" in other, ostensibly "non-religious" fields. A positive case for this embeddedness would surely require us to show how the contents of those other fields were somehow colored or materially changed by the subliminal presence of religion within them. But as yet no effort to formulate such a case has been made. In the meantime, historians continue to treat the political, the social, and the economic as if they were purely secular fields of experience, wholly unaffected by the presence of any embedded religion.

Then there are the more philosophical problems raised by claims for the embeddedness of Athenian religion. Most immediately, if the Athenians did not in fact divide their world into discrete functional fields, like the political, the social, and the economic in the first place, what precisely was the nature of the fabric that religion was embedded in at the time? And if this embedded religion had no discernible causal or constitutive effect upon the societal fabric in which it was embedded, in what sense exactly can we say it was even there at all? How exactly could it be "in" anything as something other than itself? Again, if we are not prepared to do the hard work of specifying religion's quiddity as an "embedded" phenomenon, then this whole proposition becomes little more than an impressive-sounding way of evading the issue altogether, a rather banal claim that religion must have been everywhere in Athenian reality because we cannot find it anywhere as such, as a pure, irreducible essence or thing-in-itself.[18]

Why So Many Problems?

To summarize the arguments of the last two chapters, the closer we scrutinize the consensus account of "democratic Athens," the gravitational center of the modern field of Greek history, the more problematic it begins to appear. When we historicize Athenian *demokratia* as an ancient ancestor of our own liberal democracy, we end up producing a strangely self-defeating lifeworld, one riven with all manner of improbable dissonances, paradoxes, and contradictions. We end up constructing a *polis* that is continually at odds with the "democratic" essence of its own being. And our problems only deepen when we then go in search of the specific environmental conditions that might have sustained an egalitarian, proto-liberal way of life in Athens in the first place, such as discrete fields of state, society, economy, and religion, public and private spheres, sacred and secular realms, and individualist notions of rights, freedom, and citizenship. It is not just the case that we have no evidence, explicit or implicit, for any of these phenomena as such. But surveying the evidence we do have, it is very hard to see how any of these phenomena would have made any meaningful sense in ancient Athens at the time. So what is the ultimate cause or source of all these problems?

In what follows, I would like to suggest that the ultimate problem here has nothing to do with inadequacies of evidence or data-gathering techniques. And it has nothing to do with the particular approaches or methodologies used by any particular scholars. It has everything to do with that universal model or template of social being that all current approaches and methodologies take entirely for granted. In other words, it has everything to do with the basic analytical devices that our discipline expects us to use when we historicize past experiences. To help us better understand the nature of this problem and formulate more effective alternatives, we must now expand the horizons of the enquiry a little further.

4

Historicism and Its Consequences

A Postcolonial Critique

Arguably the most powerful critique of established historical practices in recent times has come from postcolonial historians and theorists. As noted earlier, this critique is not directed at any specific set of analytical blindspots, methodological limitations, or theoretical naïveties. Rather, its ultimate target is historicism in general. As Chakrabarty and others have demonstrated, the standard universalist models and "protocols" that we use to transform data into "histories" are all premised upon a distinctly Eurocentrist way of knowing experience. In their efforts to produce a kind of cumulative biography of our species, to establish commensurabilities and continuities between all past lifeworlds, these devices fundamentally alter non-western experiences, "translating" them all into modern western terms.[1]

When viewed from "outside" in this way, historicism is thus a mode of knowledge production that allows us only to tell certain very specific kinds of stories about the past, stories that will always be grounded in Europe's peculiarly secular, "scientific" understandings of cause and effect, time and space, agency and subjectivity, truth and reason, and so on. And this means that we cannot tell meaningful stories about experiences in lifeworlds where, say, there was no self-evident distinction between orders of nature and culture, where human individuals were not the only possible subjects and actors, where deceased ancestors continued to influence events from beyond the grave, where gods and other "supernatural" forces controlled the very conditions of existence, where life itself was not bound by secular laws of science. For when we try to historicize such lifeworlds in the conventional manner, we will always end up infantilizing them as "pre-political," "pre-capitalist," "pre-scientific," or generally "pre-modern," defining them by what they have not yet but, it seems, inevitably will become. In the process, the heterogeneous truths and rationalities that once animated the essential fabrics of those worlds, making them whatever they really once were, are quite literally lost in translation.

In sum, our discipline lacks the wherewithal to recover and convey the past's myriad non-European ways of being human. Instead, it produces "histories" of non-western worlds that are no more than variations on "the history of Europe." And such histories will inevitably be full of inadequacies, incompletenesses, and failures.

Unless one is seriously willing to defend a mode of history-making that has no way to express the real, lived experiences of the past's non-western peoples, this postcolonial critique of conventional historicism would seem to be incontestible. Indeed, if there is a concern here, it is that this critique does not go far enough. For a case can be made that the kinds of problems we have identified in the last two chapters, problems in modern accounts of a proverbially "European" life-world, likewise originate in the very foundations of our disciplinary practice. The analytical vicissitudes of our "democratic Athens," I suggest, are also products of conventional historicism. More specifically, they are products of that universal analytical template, the general model of social being which allows us to turn data into histories in the first place.

To press the case further, this chapter will look at some of the more specific analytical problems that arise when we insist on translating a non-modern *demokratia* into a quasi-modern or proto-modern "democracy." We can begin by looking at how this act of historicist translation shapes the kinds of the stories we tell about the formation of the Athenian *politeia*. In what follows, we shall consider three widely cited accounts of this process, each of which quite explicitly historicizes the evolution of *demokratia* as a modern-style secular process of "democratization," a progressive extension of a political or civic equality to every (male) Athenian.

Stories of "Democratization"

First, in Ian Morris's much discussed account, "the Greek *polis*" is seen as an enterprise that was stamped from the start by a distinctive "middling ideology," one that prescribed an "equal share in the community" for every "citizen." For much of the archaic period, Morris contends, this dominant ideology was challenged by a rival "elitist ideology" that constituted rich aristocrats as the natural, exclusive rulers of every *polis* due to their closer relations to "external sources of authority," like gods, Near Eastern kings, and the heroes of the past. Morris then explicitly enlists the ideas of Robert Dahl, a prominent theorist of modern liberal democracy, to argue that a shift towards formal, secular "democracies" in Athens and elsewhere was finally made possible by a "crystallization" of a "Strong Principle of

Equality" in the period ca. 525–490 BC, a process that apparently precipitated the "collapse" of the elitist ideology. Thus "powerless in the face of growing citizen confidence, aristocrats everywhere conceded the second proposition in Dahl's Strong Principle of Equality, that no external source of authority made them so much better qualified than other citizens that they alone should automatically be entrusted with making the collective and binding decisions."[2]

No less explicitly liberal and modernist is Josiah Ober's claim that the "birth of democracy" in Athens in 508/7 BC was stimulated by a "popular revolution." Against conventional "constitutionalist," "top-down" accounts, which credit the shift to *demokratia* in Athens to the reforms of the aristocratic Cleisthenes, Ober expressly compares this transformation to events in France in 1789. According to later sources, when Cleisthenes' rival Isagoras summoned his friend king Cleomenes and a force of Spartans in an attempt to force Cleisthenes out of Athens and prevent the implementation of the reforms, many Athenians besieged them on the acropolis and subsequently drove them from the city, thereby allowing Cleisthenes to return and his reform program to proceed. While the sources tell us little about this event and nothing at all about the identities of the Athenians involved, Ober reads it as a "leaderless riot" by a "socially diverse" body of citizens, one that broadly resembled the "mass uprising" which led to the storming of the Bastille in 1789.[3] As such, this revolution "crystallized" an "epistemic shift," one that rendered participation by poorer Athenians ("thetes") in government duly conceivable. In short, it paved the way for a new "democratic" regime, sustained by an "ideological hegemony of the masses."[4]

Third, we have the account of Kurt Raaflaub, which is framed as a critical response to Ober's story of a late sixth-century "revolution." Like Ober and other specialists, Raaflaub assumes that "democracy" in Athens was defined ultimately by its egalitarian commitment to the "full participation and sharing of power by the lower classes." But as he reads the evidence, this condition was not met until 462/1 BC, when the reforms of Ephialtes somehow "weakened" the Areopagus council, an archaic, elite-dominated body that had apparently continued to exercise some largely unspecified "powers" down to this time. If we then want to explain why political participation by thetes became routine at this particular juncture, Raaflaub would direct our attention to the significance of the Athenian navy, a force manned primarily by the poor, which had been dramatically expanded in the late 480s and had since enabled Athens to exercise a growing Aegean-wide hegemony. In sum, while the reforms of 508/7 may well have "enhanced equality and participation" in a general way, democracy "in the full sense of the word" was not realized in Athens until nearly fifty years later.[5]

The differences between these three stories do not much concern us here. For our immediate purposes, what matters is what they all have in common. All three of these accounts simply take it for granted that ancient *demokratia*, like modern democracy, was premised upon egalitarian principles, upon the need to express and institutionally recognize an apparently innate equality among individuals. They thus take it for granted that any story about the formation of the Athenian *politeia* must, by definition, be a story about "democratization," about the progressive extension of equal political rights to all free, native-born males in the *polis*. But one is unlikely to find any explicit, incontrovertible support for this assumption in the Athenian evidence, for at least three reasons.

First, the stories that Greeks themselves told about the origins of *demokratia*, stories that are recorded or presupposed by all our sources, bear little resemblance to our own. As noted earlier, the classical Athenians believed that their *politeia* was a primordial ancestral way of life. Aside from the brief interruption caused by the personal rule of the Peisistratid "tyrants" (ca. 546–510 BC), it had prevailed more or less continually in Attica since the days of the early kings, at least since the time of king Theseus. Thus, in Athenian eyes, while later "lawgivers" like Solon, Cleisthenes, and Ephialtes may have modified certain institutions or added new ones along the way, these interventions involved mere adjustments to an existing way of life, not attempts to create a new one. And even in our more "reliable," more "scientific" sources, like Thucydides' *History* and the Aristotelian *Constitution of the Athenians*, one finds nothing which directly contradicts the general truth of this same basic storyline.[6] There simply was no memory of any growing agitation by non-elites for "equal rights" in Athens or anywhere else in (what we would call) the late archaic era. There was no memory of any dramatic historical "rupture," never mind any "popular revolution," in Athens in 508/7 BC. Nor was there memory of any landmark "birth of democracy" in the late 460s.[7]

No less revealing, the one transformative event in the life of their *politeia* which the Athenians did actively commemorate was an event that had nothing at all to do with Cleisthenes, Ephialtes, "the masses," or proto-liberal political innovations of any kind. The incident in question took place several years before Cleisthenes' reforms in around 514 BC. At first sight, it involved nothing more than the killing of one elite Athenian by two others, who were themselves killed in the aftermath. But what gave the event its significance was the identity of the victim, namely Hipparchus, a son of the "tyrant" Peisistratus. All evidence indicates that the killers, the hitherto obscure Harmodios and Aristogeiton, were thus celebrated soon afterwards as "Tyrannicides" (*turannoktonoi*) who had liberated the people of Athens from a period of oppressive rule by a single family. In due course, the *demos* established annual cult honors for the pair, elevating them to the status of "heroes" of the *polis*. As a visual memorial of their extraordinary achievement, imposing life-sized statues of the tyrant-slayers were also erected in a prominent place in the agora, the center of social life in the *polis*. And over time, memory of the event entered popular lore, as a subject of drinking songs and a theme for vase painters.

This, then, was the closest we come to a transformative, "Bastille Day" moment in Athenian memory. And as all of our sources consistently assume, state, or imply, the celebration of the "Tyrannicides" as such makes sense only if it was generally accepted that their actions made possible the restoration of a traditional, ancestral *politeia*, not the creation of a new one. Accordingly, these same sources will have nothing meaningful to tell us about the "birth" of any proto-liberal "democracy" in the later sixth or earlier fifth century.[8]

Second, evidence for the very existence of any such proto-liberal "democracy" is no more compelling or unambiguous after ca. 460 BC, when we have a far richer trove of sources to work with. To be sure, if one takes a rather selective view of the Athenian *politeia*, focusing purely on its "political" contents, one can choose to see distinctively egalitarian principles at work in certain specific practices and procedures, such as the payment of small stipends to many office-holders and the widespread use of a lottery to appoint most of those same officials. But as we shall see later in the book, claims that such mechanisms were animated by a liberal-style individualist ethos of equality become a good deal less compelling when we reconsider them in their broader context, as components of an entire "way of life." More immediately, numerous other components of that same *politeia* should encourage us to question whether any such ethos was really there at all.

To begin with, one has to note that the vast majority of the stipendiary, lotteried offices in Athens conferred little or no significant power or authority on individuals. Rather, they were almost always collegiate and purely functionary in nature.[9] At the same time, positions of more urgent, potentially life-and-death significance, positions which did require holders to possess some prior level of knowledge or experience, like most treasurers, the ten generals, cavalry commanders, warship designers, and overseers of the water supply, were invariably filled by election and thus typically held by members of the wealthy, relatively educated minority.[10] Likewise, for reasons noted earlier, members of this same tiny minority all but monopolized the role of assembly "advisor" and tended to be visibly overrepresented in the ritual and ceremonial life of the *polis*.

As for the thetes, as we also noted earlier, there is no evidence at all that they ever constituted themselves as a free-standing, organized group, one that was capable of formulating its own distinct identity and interests, promoting its own "leaders," or mounting a concerted action of any kind. As an identifiable group or class, they had no conspicuous presence in *polis* rituals and ceremonies. Even as rowers in the all-important navy, they are all but missing from Athenian military monuments and commemorations, whose imagery typically represents cavalrymen and infantrymen. They almost certainly did not take part in the *ephebeia*, the closest thing in Athens to a male rite of passage from youth to adulthood, which offered training only for would-be hoplites. And there is even some doubt as to whether their names were always entered in the records of member enrollments that were maintained by each deme.

Surveying such evidence, Raaflaub himself elsewhere concludes that "there may have been noticeable discrepancies between the ideology and the reality of democratic equality—discrepancies that correspond to widespread and well-known aristocratic tendencies to exclude and denigrate the despised members of the lower classes."[11] But if these "aristocratic tendencies" were still so visible and so unrestrained in all areas of Athenian life, from assembly leadership and deme enrollments to military and ritual activities, can we really still call them "discrepancies"? If the "ideology of democratic equality" was so consistently violated by Athenian "reality," what exactly is causing us to believe that any such proto-liberal ideology was really there in Athens in the first place?

This brings us to the third point, which is the complete absence of evidence from Greek thought for the very specific kinds of premises and presuppositions that any erstwhile commitment to a proto-liberal egalitarian ideology would seem to require.[12] The ancient Greeks knew no equivalents of modernity's great liberal manifestos, like the U.S. Declaration of Independence or the French Declaration of the Rights of Man and Citizen. They knew nothing of any universal, pregiven world of autonomous, rights-bearing individuals, where all are equally predisposed and equipped by nature to pursue their own "life, liberty, and property."[13] Simply stated, there was no Greek John Locke or Adam Smith, no Greek James Madison or John Stuart Mill. And in the absence of any such liberalism *avant la lettre*, I submit, it would seem to be unrealistic to expect to find evidence for the realization of any "Strong Principle of Equality" across "much of" late archaic Greece, for a French-style "revolution" in late sixth-century Athens, or for the birth of democracy "in the full sense of the word" nearly fifty years later.

Yet once we historicize *demokratia* as "democracy" to produce our vision of "democratic Athens," these are just the kinds of origins stories that we are predisposed to tell. We commit ourselves from the start to telling stories that resonate with our own modern experience of liberal democracy, even if this requires us to work constantly against the grain of our sources. In so doing, we rule out the possibility of telling potentially more interesting, perhaps more historically meaningful stories, stories about an altogether less familiar phenomenon, a non-modern, non-liberal *demokratia*.

A World Never Fully Itself

But there are other more serious consequences of historicizing *demokratia* as "democracy." Perhaps above all this act of historicist translation inevitably prejudices the way we think about Athenian practices and institutions in general. For it encourages us to use categories which implicitly measure the components of an ancient way of life by our own modern standards, standards against which non-modern practices will always fall short.

For example, we tend to see a government in Athens that was somehow manned and led by mere "civilians" or "volunteers." We conventionally see a legal system where all cases were somehow prosecuted by "private citizens" and decided by "jurors," even though the Athenians expressly called them "judges" (*dikastai*). We commonly see a system of policing that still depended largely on "informal" mechanisms of "self-help" and "social control." We see a system of religion that still depended heavily upon hereditary, "privately held" priesthoods. We see a mode of war-making which still relied ultimately upon "militias" of "citizen soldiers." And the problem with all of these standard categories is that they presuppose the possibility of a very different lifeworld, one that is ordered entirely otherwise.

For terms like "layman," "volunteer," and "civilian" only have meaning and valence in a societal environment like our own, where it is entirely possible, even normal to entrust the most urgent, life-sustaining administrative functions to impersonal apparatuses of credentialed professional specialists. Such terms have no meaning at all when applied to an environment like classical Athens, where an order run by full-time, trained experts was all but unthinkable. Worse, by using terms like "layman," "volunteer," and "civilian," all we do is create a general impression of a "simpler," almost "pre-political" polity, a polity "still" governed by "amateurs," a world that has "not yet" discovered the virtues of a modern-style rational-legal, professionalized mode of population management.

Simply put, when we historicize *demokratia* as "democracy," we inevitably end up defining the Athenian *politeia* as much by what was not there as by what was. We cannot help seeing a lifeworld that was defined largely by negatives, by withouts and not yets, by an absence of a certain modernity. We cannot help seeing a lifeworld that was suspended in a kind of perpetual infancy, forever falling short of its own full self-realization, a realization that we know all too well will come in some other, much later time, namely our own. And of course this overall impression of incompleteness is only amplified by all those "paradoxes" and "contradictions" discussed in Chapter 2, "problems" which also arise from our efforts to historicize *demokratia* as "democracy."

Thus, instead of trying to understand Athenian gender relations on their own terms, our accounts of female experience begin inevitably with the "fact" that women were systematically "denied" their "political rights" and "excluded" from the body of "full citizens." And knowledge of this basic "fact" duly predisposes us to look for further signs of "gender inequality," "patriarchy," even "misogyny" in Athenian life, profoundly coloring our reading of the evidence. In much the same way, our conviction that a "democracy" somehow coexisted with slavery and a ruthless imperialist enterprise in the same *polis* prompts questions about the ethical sincerity and maturity of the *polis* in question. It causes us to assume that the Athenians were self-servingly selective in their recognition of individual rights, that they were too unenlightened or unsophisticated to see the egregious contradiction between their exploitation of non-Athenians and their "democratic" ideals.

So too, as Raaflaub's remarks above indicate, our conventional modern understanding of *demokratia* predisposes us to see an Athens that had not yet fully abandoned the more hierarchical habits of its "pre-democratic past," an Athens where "the aristocratic tendencies to exclude and denigrate the despised members of the lower classes" were still unfortunately "widespread." As Raaflaub himself goes on to observe of "evidence" for continuing "discrepancies" between ideology and reality in Athens:

> At the very least, this evidence illustrates what obstacles, rooted in century-old traditions and prejudices, may have made it very difficult to justify and fully realize a concept so revolutionary as that of almost unlimited civic equality. Nothing short of a complete overhaul of political values and ethics and a radical reeducation of the citizen body, it might seem to us, could have overcome these obstacles.[14]

For Raaflaub, as for the field as a whole, it seems, our conviction that Athens was "democratic" is just incontestible, even if this means historicizing so many essential, integral components of the Athenian way of life as "problems," "contradictions," and "obstacles to unlimited civic equality."

In short, when we historicize *demokratia* as we do, we leave classical Athens suspended in a state of perpetual semi-enlightenment, forever in need of that "complete overhaul" of "political values and ethics" that we know will surely come many centuries later in a very different time and place. And so long as we see Athens as a sadly immature, incomplete anticipation of a modern liberal polity, we will be incapable of understanding it on its own historical terms, as a complete and fully realized version of itself.

A Suspect Device

Yet despite all these problems, our historicist vision of an Athenian "democracy" persists, largely unchallenged. And the reason it still seems to make such good sense to us, presumably, is because of the way we typically historicize the *polis* as a whole.

For whenever we use the standard, modern historicist template to objectify the broader Athenian environment, seeing it as a functionally compartmentalized terrain peopled by natural individuals, we inevitably see a lifeworld where conditions were always already congenial to the formation of a "democracy." We see a lifeworld where equalities in a discrete, proto-liberal "political" realm could potentially coexist with manifest inequalities in other, apparently less significant but no less discrete realms, like the "social" and the "economic," where "traditional," "pre-democratic" logics and principles might still have prevailed. We see a lifeworld where a modern-style "citizenship" could potentially flourish in a circumscribed "public" sphere of the *polis*, a space securely insulated from the "private" sphere of the *oikos*. And we see a lifeworld where a "sacred" realm of "religion" could potentially differentiate itself from a "profane" realm of the "state," thereby opening up a space where a secular, proto-liberal "democracy" might once have thrived.

And even if we are then prepared to recognize the lack of evidence for this standard model of social being and modify it accordingly, our practice will still be no less problematically historicist. For when we speak instead, say, of a "fusion" between Athenian state and Athenian society or of the "embeddedness" of an Athenian economy and an Athenian religion, we are not analyzing antiquity on its own ancient terms. Rather, we are merely positing a hypothetical "pre-modern," antecedent other, a kind of reverse image or binary opposite of whatever it is we have already defined as "modern." So all we are really doing is reinforcing our cumulative modernist, historicist impression of a lifeworld that is forever incomplete and inadequate. Apparently, the Athenians were still too illiberal or intellectually narrow, say, to recognize the need to secure a fully protected realm of privacy, to acknowledge the presence in their midst of a free-standing economy, or to mandate the kind of "separation of church and state" that rational, liberal progress so obviously requires. In other words, for all the demonstrable learning and sophistication of Morris, Ober, Raaflaub, and so many others in our field, the discipline's basic analytical common sense predisposes us to produce histories which can too easily look like rather hollow exercises in modern liberal self-congratulation.

To conclude, our consensus vision of a "democratic Athens" vividly illustrates the limitations of conventional historicist practice. It demonstrates quite clearly how the tools and categories of our standard analytical template condition us to tell certain very specific kinds of stories about the past, stories about just one single way of being human. In their own way, our accounts of classical Athens are thus just as fundamentally troubled and troubling as all those "histories" of, say, "India," "China," and "Kenya" which prompted the postcolonial critique of historicism in the first place.

All of which should in turn encourage us to recognize that historicism's limitations are not a function of its Eurocentrism or its "western-ness" as such. They are ultimately a function of its modernism. For it is the modernism of our historicism that allows us to presume a genuinely scientific, panoptic knowledge of the past, a knowledge superior to any possessed by all those peoples who actually lived that past. And it is the modernism of our historicism which thus authorizes us to homogenize experiences in all non-modern lifeworlds, western and eastern alike, along post-Enlightenment, European lines. But by appointing us the god-like arbiters of all historical truth in this way, this historicism ends up defeating its own high-minded aspirations. Instead of enabling us to recover past ways of life in all their variegated multiplicity and particularity, it again causes us to lose them in translation.

5

Beyond Cultural History

HOW, THEN, CAN we hope to retrieve something more of that other Athens, the Athens which was actually lived and experienced by the Athenians themselves? What would their celebrated *politeia* look like if we no longer viewed it through the lens of conventional historicism, using our standard modern analytical template? How different does their *demokratia* appear when we try to see it more on their terms, in the world of their experience?

At first sight, there would appear to be three possible approaches available to historians who want to pose and answer such questions.

The Dualist Bind

The first and most obvious would be to deploy the methods of mainstream cultural history. For more than thirty years now historians have been expressly exploring the mindscapes, mentalities, ideologies, and imaginaries of their subjects.[1] And there can be no doubt that this turn in our practice has helped to sensitize us to the alterities of past experiences. No doubt, it has allowed us to take seriously non-modern beliefs as objects of study in their own right, helping us to see that past peoples experienced realities quite different from our own. But there is one kind of non-modern belief that cultural history never allows us to take seriously, and it is the most important one of all. This is the general belief that gods, magical forces, and other alien contents of past experiences are not actually figments of "belief" at all, but real, independently existing phenomena. And the reason we cannot take this particular belief seriously is because cultural history, like materialist history, is in the end a product of "Europe's" peculiarly materialist way of knowing all experience, a knowledge whose standards of truth and realness are determined by modern, western laws of physics and nature.

For all its well-intentioned interest in the alterities of past lifeworlds, cultural history presupposes a kind of primordial divide between matter and meaning, between being and knowing, between a pregiven, material reality and the "culture" one uses to represent or make sense of that reality. Like traditional materialist histories, it remains committed to a Cartesian dualist metaphysics, whereby experience is dichotomized into objectively real material phenomena on the one hand and merely subjective, ideational phenomena on the other, each of which can then be presumed to have their own separate histories. The very idea of a history of mental representations makes sense only if one presumes that there is always already a materially self-evident, mind-independent reality "out there" to be represented, whether it is seen from the "emic" (insider) perspectives of our subjects, or from the altogether more scientifically informed, god's-eye or "etic" (outsider) perspective of ourselves.

In short, mainstream cultural history takes for granted the very same universal analytical template as materialist history, a template which would have us dichotomize all the contents of past experience into material and cultural phenomena in the first place. Far from threatening or disturbing historicism's philosophical foundations, cultural history at once presupposes and reproduces those very same foundations. It is simply historicism's way of translating data that cannot be accommodated within standard "etic," materialist narratives.[2]

Thus, even when our non-modern subjects insist that, say, gods and other superhuman powers shaped the fortunes of their lifeworlds, our historicism cannot accept the truth of such experiences. As Chakrabarty has observed, the best it can do is to "anthropologize" phenomena that defy our modern scientific understanding of "the world," reducing them to mere ideas, beliefs, and representations, translating them into "culture."[3] And such a practice would seem to be ethically troubling, to say the very least. For it presumes that we moderns must already possess a true knowledge of this same apparently mind-independent world, a knowledge that is somehow not itself just a figment of our own prevailing beliefs or culture. It presumes that we can somehow know the contents of non-modern experiences better than those who have actually lived those experiences. In the words of the anthropologist Martin Holbraad, it presumes that non-modern peoples continually "get stuff wrong."[4] Conventional cultural history, just like traditional materialist history, thus ensures that we moderns remain the god-like arbiters of all historical truth.

All of which may then encourage us to consider the second possible alternative, which is a somewhat more radical form of cultural history, one that might appear to present a more serious challenge to historicism's materialist foundations. I refer to what is sometimes called "discursive history," which, in its most uncompromising forms, expressly challenges the proposition that an objectively knowable, materially self-evident past is always already there to be represented.[5] Instead, it posits a world where encultured, historically contingent "discourses," the ideational conditions for prevailing patterns of thought, speech, and action, continually and actively construct the truths of experience at any given time, rather than passively reflecting and representing those same truths. So at least in theory, discursive history would oblige us to analytically prioritize our subjects' constructions of their realities over any modern historicist accounts, since dominant discourses would presumably be among the essential fabrics from which lifeworlds are actually made. And in so doing, this practice would potentially restore to non-modern peoples, western and non-western, some power to determine the truth of their own experiences.

Why is it then that few historians so far seem willing to commit themselves fully to the cause of discursive history? Why is it that the promise of this alternative practice has been at best only selectively pursued?

For many, apparently, the implicit denial of an objectively knowable historical truth is simply too epistemologically troubling to accept, portending as it does a possible chaos of "relativism," a history where "anything goes."[6] More serious, though, are the ontological problems that are raised by all forms of discursive constructivist explanation. For if taken to its limit, this constructivism will come dangerously close to a kind of Berkeleian subjective idealism. It is one thing to suggest that what we may think of as relatively immaterial or ideological phenomena, like masculinity, ethnicity, and nationality, are cultural or discursive "constructions." But how exactly is it that pure thought can conjure into existence entire historical worlds, complete with all their hard-surfaced, physical contents? Is materiality merely some inert substrate of experience that passively awaits transformation into real, socially meaningful objects by the demiurgic powers of human culture? A pure discursive constructivism, it seems, would give us no means to explain how exactly "matter comes to matter," how materialities actively constrain and condition the production of dominant discourses.[7] To be sure, when one surveys evidence from across time and space, one will find that humans construct phenomena like weather systems, landscapes, human physiologies, even life itself in any number of different ways. But are they really free to make of these things whatever they want?

Furthermore, it still remains unclear how discursive constructivism could form the basis for any historical practice without radically reconfiguring the very philosophical foundations of history itself as an enterprise. If one is committed to an anti-materialist account of realness, to something like the proposition that "all the world's a text," one must surely question, if not outright reject, most if not all of the premises of conventional historicism, including its Eurocentrist, materialist constructions of, say, time and space, cause and agency, truth and realness, and of course the discursive and the extra-discursive.[8] But as yet, the wider disciplinary implications of this alternative practice remain under-specified. In the meantime, pursuing a selective discursive history, a history which grants full realness to, say, the contents of non-modern "ideologies" without first disavowing its commitment to the discipline's materialist grand narratives, looks a little bit too much like having it both ways. It looks like playing the game by two different sets of philosophical rules at the same time.

What then of our third alternative, which has been expressly conceived as a corrective to the discipline's ever-growing preoccupation with culture over the last thirty years? Calls for some kind of "material turn" or "new materialism" in historical analysis have been growing since the turn of the century. And again this general approach may seem promising at first sight. There is perhaps a certain immediate appeal in a history that would treat material and cultural phenomena as equally significant and equally real, one that would somehow allow us to stiffen the more theoretically sophisticated procedures of cultural analysis with a renewed sense of the irreducible materialities of experience.[9]

That said, the pioneers of this kind of approach have almost invariably been specialists in the history of western modernity. Its value to students of non-modern lifeworlds seems more questionable. A historian of, say, modern Europe can reasonably assume that the peoples she studies will share her common sense ideas about how one distinguishes an objective material phenomenon from a subjective cultural phenomenon in the first place, about where one ends and the other begins. But historians of non-modernity can make no such easy assumption.[10]

For example, to the modern eye, it may be self-evident that, say, the "Athenian economy" was an objectively real, material entity, while the goddess Athena was no less obviously a mere figment of "faith" or "belief." But to an actual inhabitant of the Athenian lifeworld, something like the reverse would have been the case. The fact that the gods of Athens were not readily visible did not for a moment nullify the truth that they were fully material agencies, the effects of whose actions were clear enough for all to see, in weather patterns, crop yields, bodily health, and so forth. By the same token, far from being a materially self-evident phenomenon, an "Athenian economy" would have made little sense at all to a classical Athenian. At best, it would have been merely an idea, and a very strange one at that.

In sum, there are problems with all of the three kinds of approaches that are currently available to those who seek an alternative mode of practice, one that is better equipped to retrieve and represent non-modern ways of being human. And in all three cases, the problems are ultimately of an ontological nature, in that they concern what we take to be the essential contents of experience, the a priori fabrics of being itself. All three seek to expand the limits of historicism by rethinking the nature and relative significance of materiality and culture in the production of the real. But in so doing, all three still take for granted a post-Enlightenment dualism which would dichotomize the world of experience into material phenomena and cultural phenomena in the first place. As a result, all three still commit us to an account of realness which would inscribe some form of modernist mind/matter and/or nature/culture divide upon all non-modern pasts. In other words, like traditional materialist history, all three still to a point take for granted historicism's standard analytical template. All three still require us to translate past ways of being human, to conform them all to our own modern ontological commitments, to our own modern presuppositions about the essential components of experience. Which is to say, all three in some way require us to superimpose peculiarly modern metaphysical conditions upon non-modern lifeworlds, thereby fundamentally altering what could and could not be really there at the time.

We are thus presented with a paradox of a far-reaching kind. Simply put, the designated mission of history as a discipline is consistently thwarted by the very same metaphysical presuppositions which made possible its formation as "scientific" enterprise in the first place. As Philippe Descola, a leading ethnographer of indigenous Amazonian peoples, has recently observed, precisely the same paradox arises in anthropology:

> [The discipline's] role is to gain an understanding of how peoples who do not share our cosmology came to invent for themselves realities that are distinct from our own, thereby manifesting a creativity that cannot be judged according to the criteria of our own accomplishments. And this is something that anthropology cannot do so long as it takes our reality for granted as a universal fact of experience, along with our ways of identifying discontinuities and discerning constant relationships in the world and our manner of distributing entities and phenomena, processes and modes of action, in categories thought to be predetermined by the texture and structure of things.[11]

And the inevitable net result is a practice that occludes non-modern ways of being human, as Descola goes on to explain:

> By turning modern dualism into the standard for all world systems, we are forced into a kind of well-meaning cannibalism, as we repeatedly incorporate nonmoderns' objectivization of themselves into our objectivization of ourselves. . . . [In so doing, we transform those peoples into] preliminary sketches of citizens, protonaturalists, quasi historians, and nascent

economists: in short, precursors who fumble at a way of apprehending things and human beings that we ourselves are believed to have discovered and codified better than anyone else. Of course, that is one way of expressing respect for them, but amalgamating them into the categories to which we belong is the surest way of wiping out their distinctive contribution to the intelligibility of the human condition.[12]

How then are we to escape this dualist bind? As I will now suggest, the problem, by its very nature, suggests its own solution.

A Possible Alternative

If the limitations of conventional historicist practices are ultimately ontological, in that they require us to superimpose modern conditions of being upon all non-modern lifeworlds, the possibility of an alternative mode of history-making duly becomes imaginable. For it follows that any more viable, less problematic way of accounting for past experiences would have to be one that, for all analytical purposes, is willing to suspend our own modern standards of truth and realness. It would have to be a mode of analysis that seeks expressly to make sense of each non-modern lifeworld on its own ontological terms, in its own metaphysical environment, as a more or less self-realizing world in its own right. In short, to avoid losing extinct lifeworlds in translation, we need to take an "ontological turn" in our practice.

Obviously, this extension of our analytical horizons all the way down to the ontological level would entail a basic change in the rules of historical engagement, a significant paradigm shift in our discipline. In the end, it would mean re-engineering historicism itself to recognize a past that was experienced in many different worlds, not just in one. And doubtless, this proposition will seem to some to be altogether too radical or too drastic at first sight. But it is essential to emphasize up front that this kind of shift would not be without intellectual parallel or precedent. On the contrary, it would share certain affinities and consonances with influential recent currents in a number of other disciplines.

In philosophy itself, for example, interest in ontology has been substantially re-energized since the millennium by the emergence of "speculative realism," an orientation associated with figures like Graham Harman, Ray Brassier, and Quentin Meillassoux.[13] And in the province of science and technology studies, a long-standing disciplinary concern with ontological questions has become significantly more explicit and systematic in recent years in the works of Bruno Latour, Andrew Pickering, and others.[14] Most germane to our purposes, though, would be cognate developments in anthropology, where Philippe Descola is far from the only specialist to recognize the metaphysical constraints imposed by culturalist and constructivist forms of analysis. Likewise motivated by the need for a practice that can produce more ethically and philosophically defensible accounts of human alterities, calls for an ontological turn in that discipline have already been the subject of intense conversations for more than a decade.[15]

Again, interest in ontological issues is not something new in anthropology. Anticipations of the proposed turn can to a point be seen, for example, in Descola's earlier work *In the Society of Nature* (1994), which revived the term "animism" to refer to the extension of personhood to non-humans by the Amazonian Achuar, whose world defies our modern nature-culture binary. Likewise, Marilyn Strathern's *The Gender of the Gift* (1989) had already explored the radically non-modern forms of personhood and gender that prevail among the Hagen of upland Papua New Guinea. Other antecedents can be found earlier still.[16] But concerted, self-conscious efforts to establish "the ontological" as a routine horizon of enquiry are rather more recent. And very much at the vanguard of this particular cause is Eduardo Viveiros de Castro, a Brazilian ethnographer of indigenous Amazonian peoples.

Drawing on ethnographic data from a range of Amazonian populations, Viveiros de Castro presses the case for the radical alterity of indigenous ontologies considerably further than Descola had done in his 1994 study of the Achuar. It is not just the case that these ontologies resist our own nature-culture binary. They effectively invert it and turn it inside out. While in our reality multiple cultures co-exist within a single order of nature, in some Amazonian realities multiple natures coexist within a single cultural order. Thus, while most if not all creatures of the forest, both human and non-human, are persons, beings who share a single common way of seeing and apprehending things, the things they actually see are different. Far from seeing any unitary, continuous "nature," each group of creature-persons inhabits what amounts to its own natural order, with its own distinct ontology, such that humans, for example, are at once jaguar-like predators in the world of the pigs and pig-like prey in the world of the jaguars.

For Viveiros de Castro, this radically non-modern "multinaturalist" metaphysics poses formidable challenges to the established "multiculturalist" foundations of anthropology, encouraging us instead to specify and represent the ontological variabilities that underwrite human experiences. And a practice along these lines would also invigorate the discipline as a whole with a new, more charged sense of ethical and political purpose. For if, as he proposes, anthropology is to be "the science of the ontological self-determination of the world's peoples," it would then be "a political science in the fullest sense."[17]

Viveiros de Castro's call for an ontological turn in anthropology has been most visibly supported and amplified by Martin Holbraad, a specialist in the study of *Ifá*, an Afro-Cuban cult practice. And like Viveiros de Castro, Holbraad uses the claims of his ethnographic subjects, the diviner-initiates (*babalawos*) of *Ifá*, to build his theoretical case for the proposed shift. In particular, he focuses on the claim made by these *babalawos* that a certain form of powder (*aché*) "is" their divinatory power. The standard question for the ethnographer raised by such a claim would of course be something like: Why is it that *Ifá* diviner-initiates *believe* that powder is power? And a standard kind of answer would be that the material powder must somehow represent the immaterial power metaphorically, since the ethnographer already *knows* that they are two different things. But as Holbraad emphasizes, this presumption of prior knowledge about the nature of powder and power in *Ifá* causes the ethnographer to misrepresent the claim of his subjects, who *know* that the two are in fact the same thing. So the right question to ask in this case would instead be something along the lines of: What would the world have to be like for powder to *be* power? As Holbraad himself puts it:

> [T]he difference between the anthropological analyst and the ethnographic subject lies not in the different perspectives each may take upon the world (their respective "world-views" or even "cultures") but rather in the ways in which either of them may come to define what may count as a world, along with its various constituents, in the first place.[18]

So again, as in the work of Viveiros de Castro, the ethnographic encounter invites one to consider a new horizon of ontological plurality, seeing radical alterities as products of more or less self-contained "different worlds," not merely of "different world-views." Again, this new horizon demands a new form of practice, a distinctly "recursive" practice that would analyze each different human world on its own particular ontological terms. Again, too, the ethnographic data themselves provide helpful theoretical tools for justifying such a practice. For in the powder-power of *Ifá* itself, Holbraad would find an implicit challenge to our common sense dualist metaphysics, a challenge which compels us to recognize that "concepts are real and reality is conceptual."[19] And again, like Viveiros de Castro, Holbraad is ultimately most excited by the wider critical and political ramifications of an ontological anthropology. Just as the diviner-initiates of *Ifá* can to a point manipulate the powder-power of their cult to reshape the inherently mutable or "motile" truths of their existence, so too anthropologists can draw on their encounters with radical alterity to inventively define, or "infine," the new thing-concepts that might be realized in alternative future worlds.[20]

Finally, one should mention the rather different contribution made to the larger cause of ontological anthropology by Descola himself in his most recent major work, *Beyond Nature and Culture* (2013). Whereas the ontological turn of Viveiros de Castro and Holbraad seems to be motivated ultimately by a will to use ethnography for non-anthropological purposes, above all to challenge the metaphysical certainties of our modern capitalist conjuncture and to formulate possible alternatives, the principal aim of *Beyond Nature and Culture* is more narrowly intra-disciplinary. Essentially, the work is a grandly ambitious exercise in neo-Lévi-Straussian ethnology, one that would potentially classify all historical peoples according to the different ways they objectify relations between the human self and the non-human other. As Descola sees it, human cognition allows for four basic possibilities, four primary "modes of identification" or "types of ontology."

In the "animist" worlds of certain indigenous peoples in Amazonia, the circumpolar regions, and Southeast Asia, humans and non-humans possess similar "interiorities" (selfhood, subjectivity, intentionality, etc.) but fundamentally different "exteriorities" (bodily forms).[21] In the "totemist" ontologies that are characteristic of aboriginal peoples in Australia, a human collective will share both a similar internal disposition and a similar physicality with its emblematic non-human totem, which may be, say, an eagle or a kangaroo.[22] Far more common historically is the "analogist" mode of identification, whereby humans experience the cosmos initially as an infinitude of disparate, discontinuous entities, sharing no internal or external characteristics with any other beings at all. To impose some sort of order upon this chaos, they duly identify meaningful relations or patterns of correspondence between phenomena, like "great chains of being" or analogical links between a celestial realm of divinity and a terrestrial realm of humanity.[23] And then there is the "naturalist" type of ontology, which is peculiar to western modernity. This fourth type is effectively the inverse of the animist type in that it posits substantial, "natural" continuities between human and non-human physicalities, but radical discontinuities between their respective interiorities. Thus, the proposition that there exists a unitary, free-standing order of "nature," a purely material dispensation that is somehow external to a subjective human self, is imaginable only within this characteristically modern type of ontology.[24]

Predictably, these bold, paradigm-shifting proposals have been a source of considerable controversy in anthropology, provoking a mixed, sometimes heated response from specialists. A growing number have enthusiastically embraced the larger cause of an ontological turn and have sought to extend it in various ways, even if they might at times qualify or question some of the particular claims made by its original advocates.[25] As for the more vocal critics, some seem to misunderstand the nature of the arguments made for the shift. Others simply reject these arguments outright, because they cannot accept that there are serious problems with how we moderns conventionally define what counts as a world, problems to which an ontological turn would be an effective solution. And still others are probably just threatened by so radical a challenge to anthropology's philosophical foundations, not least because it seems to put at risk the very unity and universality of *anthropos*, the species-figure that is the discipline's ostensible object of study.

Whether or not these particular concerns are fair or legitimate, it is clear that only a minority of specialists have so far been persuaded to accept the proposition of an ontological anthropology. I would contend that this outcome has less to do with the ontological turn's inherent virtues than with the strength of the arguments that have so far been made on its behalf. Three weaknesses in the case seem to be particularly evident.

Consolidating the Case

First and most obvious, the case for a "many worlds" anthropology has to date been built very largely upon a small, rather selective inventory of ethnographic accounts, especially upon studies of a few specific worlds in Amazonia and elsewhere. As philosophically suggestive as this particular sample may be, it is simply too narrow to bear almost the full the weight of the case for such a far-reaching paradigm shift. The case would be significantly more compelling if it were to adduce evidence from a broader, more representative range of non-modern worlds, including those that are conventionally studied by historians, thereby further highlighting the exoticism of our own modern metaphysical and ontological commitments. Otherwise, as things currently stand, it is all too easy to dismiss the ontological alterities that are instantiated in, say, Afro-Cuban *Ifá* practices, Amazonian "multinaturalism," or animist worlds in general as the exotic anomalies, as exceptions to more or less universal rules.

Second, for all the critique of philosophical orthodoxies that one finds in works by champions of the proposed paradigm shift, the ontological particulars of these orthodoxies remain strangely under-specified. Too often they are reduced to a rather vague and over-generalized nature-culture binarism or mind-matter dualism, which is then understood to stand for a total "system" of "Western" or "Euro-American" metaphysics. In order to formulate an effective alternative mode of practice, a practice that would not automatically impose peculiarly modern ontological presuppositions upon non-modern worlds, one must first develop a more complex and nuanced sense of how we moderns generally define what counts as a world. And one must then consider more precisely how this definition shapes and informs our standard analytical devices, perhaps especially that universal template of social being discussed earlier. After all, neither "naturalism" nor dualism can in themselves explain why we specifically imagine all lifeworlds, past and present, as functionally divisible arenas or terrains that are inhabited by natural, presocial individuals.

But it is the third of these three problems that is probably the most fundamental and consequential. As Paolo Heywood has put it, the champions of the ontological turn "neglect to acknowledge that insisting on the 'reality' of multiple worlds commits you to a meta-ontology in which such worlds exist."[26] In other words, they fail to provide an adequate answer to the question: What would the metaphysical environment have to be like for it to enable the realization of many different real worlds, all with their own distinctive ontologies?

In all probability, Viveiros de Castro and Holbraad would simply reject the premises of this challenge. From their perspective, committing to a particular meta-ontology would mean committing to a particular ensemble of metaphysical essences and foundations, which would largely defeat the point of the ontological turn as they conceive it. To account for worlds where "concepts are real and reality is conceptual," where ontological truths are inherently "motile," they would instead recommend embracing something like the more open-ended "virtual" metaphysics of Gilles Deleuze, where relations effectively precede entities, where differences produce identities rather than the other way round.[27] In this alternative account, being itself would then no longer be a given state but a more or less unbounded process of perpetual becoming, a kind of eternal cosmic flux. As such, it would not be constrained, governed, or defined by the kind of fixed a priori conditions that the term "meta-ontology" would seem to demand.

Yet a response along these lines seems only to raise further questions. Committing to such a radically fluid, anti-essentialist, anti-foundationalist vision of reality may be convenient for those who ultimately desire a more or less unencumbered freedom to "infine" the contents of possible future worlds. But it is rather less helpful to those of us who are more immediately concerned with making sense of the very specific worlds that have actually been lived and experienced as real by particular peoples of the past and the present.

While a practice that takes for granted modernity's scientific "laws" of being, both natural and social, is too inflexible to accommodate the possibility of ontological variabilities across time and space, a practice that is premised on a Deleuzian virtual metaphysics would seem to veer to the opposite extreme, imposing a kind of "ontological anarchy" on the worlds of our subjects, as Viveiros de Castro himself has cheerfully acknowledged.[28] In so doing, it would clearly struggle to do justice to the enduring, centuries-long metaphysical stability of so many historical worlds, to that most common of human convictions that the essential fabrics of one's existence are timeless, immutable, and could not be otherwise. For that matter, it would struggle to explain how our own modern, disenchanted, dualist order of things, with all its neatly compartmentalized realms of experience, remains so ineffably true and so immediately real to so many of us. And it would even struggle to explain how the "multinaturalist" worlds of some indigenous Amazonians have remained so stably and consistently "multinaturalist" for as long as anyone knows.

In short, this radically anti-essentialist, anti-foundationalist species of ontological anthropology would seem to have trouble dealing with the very "givenness" of ontological phenomena, the very quality that would most define those phenomena as "ontological" in the first place. The point of the exercise is surely to devise a practice that will allow us to recognize the enduring truth and realness of the laws, essences, and foundations of being in every stable human world, our own included. If we commit instead to a vision of ontological anarchy, where there ultimately can be no such world-sustaining regularities, how then could we speak meaningfully of distinct, bounded "worlds" at all?

No less important, an ontological anthropology that is committed to the proposition that "concepts are real and reality is conceptual" leaves itself open to the easy charge that it is not really offering anything new at all, just a somewhat more radically "idealist" version of a familiar cultural relativism.[29] To respond to this charge by simply adducing ethnographies of non-dualist worlds may be enough to satisfy Deleuzians and others who are already comfortable with more esoteric postmodernist modes of critique. But it will probably not satisfy those most liable to level such a charge in the first place, namely those who are more reluctant to abandon a modern dualist common sense. For this common sense will always insist that the alterities of the ethnographic record, no matter how radical, attest merely to different worldviews, not to different worlds altogether. So if the ontological turn is ever to become something more than just another renegade cause of a radical minority, its advocates must make a more concerted effort to challenge the orthodox commitments of their more outspoken critics. And ultimately, as Heywood implies, they must harness whatever appropriate theoretical resources are available to fashion a more compelling alternative to metaphysical dualism, a counter-metaphysics that would sustain an ontological plurality without committing us to an ontological anarchy.

In its two remaining parts, the book aims to present a less tendentious, more persuasive case for an ontological turn in the study of non-modern peoples, addressing all of the above issues along the way. As noted in the Introduction, Part II presents ethical (Chapters 6, 7) and philosophical (Chapter 8) arguments for this paradigm shift, then specifies what such a move would actually involve in practice (Chapter 9). Part III revisits the ancient Greek case study to demonstrate the practical analytical benefits of an ontological turn, showing how the shift allows us to produce a more historically meaningful account of Athenian *demokratia*.

First, to establish *prima facie* ethical grounds for the turn, we should now consider further evidence for the plenitude of different worlds that have been experienced by non-modern peoples.

PART TWO

The Many Real Worlds of the Past

There is no way out: we either call quarks and gods equally real, but tied to different circumstances, or we cease to talk about real things altogether.

PAUL FEYERABEND, *Farewell to Reason*

6

Other Ways of Being Human

ALL HUMAN COMMUNITIES make assumptions about the essential nature and ingredients of their social experience. In order to act in and upon the world as a group, they must know that world as a group. They must share a general, common sense knowledge of the basic subjects, objects, relations, and processes of which that world consists. They must broadly share an understanding of reality, an account of what is and what is not always already there. They must share a way of objectifying the phenomena that form the foundations and the essences of their common experience, phenomena like those we would call personhood and subjectivity, kinship and sociality, freedom and authority, humanity and divinity, and the sources, means, and ends of life itself. And they will accordingly premise their shared ways of acting in the world, all their practices, institutions, and organizational mechanisms, upon the realness of such foundations and essences. In other words, every community, past and present, at once takes for granted, acts upon, and thereby summons to material life its own particular ontology, its own account of what it deems to be the real world. Which is to say, there have been innumerable different worlds in history, not just one.

To substantiate the point and convey at least a preliminary sense of history's remarkable ontological variability, its many possible ways of being human, Chapter 6 considers evidence from a range of different non-modern worlds. We can begin by revisiting the work of Eduardo Viveiros de Castro, looking in a little more detail at his influential account of the "multinaturalist" ontologies of certain indigenous Amazonian peoples.

Animals Are People Too

In his seminal paper on these ontologies, Viveiros de Castro draws on "various South American ethnographies," including studies of peoples in the Vaupés region of Colombia and in the Rondonia and Middle Xingu River regions of Brazil. He introduces the essential contents of Amazonian multinaturalist worlds as follows:

> Typically, in normal conditions, humans see humans as humans, animals as animals and spirits (if they see them) as spirits; however animals (predators) and spirits see humans as animals (as prey) to the same extent that animals (as prey) see humans as spirits or as animals (predators). By the same token, animals and spirits see themselves as humans: they perceive themselves as (or become) anthropomorphic beings when they are in their own houses or villages and they experience their own habits and characteristics in the form of culture—they see their food as human food (jaguars see blood as manioc beer, vultures see the maggots on rotting flesh as grilled fish, etc.), they see their bodily attributes (fur, feathers, claws, beaks, etc.) as body decorations or cultural instruments, they see their social system as organized in the same way as human institutions are (with chiefs, shamans, ceremonies, exogamous moieties, etc.).[1]

Two of the more salient constitutive features of these Amerindian worlds shed further light on their inner logics and workings.

First and perhaps more obvious, these are worlds that are committed to a distinctly animist notion of personhood. Simply stated, they are worlds where humans are not the only persons, where animals are people too. While these various types of person may bear the external form of one species or another, this is no more than a kind of "envelope" or "clothing," a mere surface appearance of physical and behavioral characteristics that conceals a common internal form. And according to Viveiros de Castro, this shared interiority of personhood is generally understood as a "soul" or a "spirit," as an "intentionality or subjectivity" that is "formally identical to human consciousness."

> At first sight then, we have a distinction between an anthropomorphic essence of a spiritual type, common to animate beings, and a variable bodily appearance, characteristic of each individual species but which rather than being a fixed attribute is instead a changeable and removable clothing. This notion of "clothing" is one of the privileged expressions of metamorphosis—spirits, the dead and shamans who assume animal form, beasts that turn into other beasts, humans that are inadvertently turned into animals—an omnipresent process in the [worlds] proposed by Amazonian ontologies.[2]

In short, in a kind of inversion of our modern dualist common sense, "nature" in these ontologies is plural, mutable, and discontinuous between species, while "culture" is singular, constant, and continuous.

What then are the "origins" of this trans-species form of personhood? Viveiros de Castro tells us that Amerindian "myths" routinely refer to an "original state of undifferentiation" between humans and animals, whereby the "original common condition" of both was "not animality but humanity." These stories duly recount how animals later lost certain human characteristics, while humans "continue as they have always been." Thus, "animals are ex-humans, not humans ex-animals."[3] Nonetheless, animals did retain their inner condition of essential humanity, the "souls" that are the source of their continuing personhood and subjectivity. Viveiros de Castro draws the following conclusion:

> It is not that animals are subjects because they are humans in disguise, but rather that they are human because they are potential subjects. This is to say *Culture is the Subject's nature*; it is the form in which every subject experiences its own nature. Animism is not a projection of substantive human *qualities* cast onto animals, but rather expresses the logical equivalence of the reflexive *relations* that humans and animals each have to themselves: salmon are to (see) salmon as humans are to (see) humans, namely, (as) human. If, as we observed, the common condition of humans and animals is humanity not animality, this is because "humanity" is the name for the general form taken by the subject.[4]

This brings us to the second characteristic feature of Amzonian multinaturalist worlds, which is what Viveiros de Castro terms their constitutive "perspectivism." By virtue of their common subjectivity, all persons, human and non-human, see or represent the world in the same way. But "what changes is the world that they see," such that each species group in effect inhabits its own distinct world.

> Animals impose the same categories and values on reality as humans do: their worlds, like ours, revolve around hunting and fishing, cooking and fermented drinks, cross-cousins and war, initiation rituals, shamans, chiefs, spirits. . . . It could only be this way, since, being people in their own sphere, non-humans see things *as* "people" do. But the things *that* they see are different: what to us is blood, is maize beer to the jaguar; what to souls of the dead is a rotting corpse, to us is soaking manioc; what we see as a muddy waterhole, the tapirs see as a great ceremonial house.[5]

Thus, whereas a modern dualist metaphysics would allow for multiple different ways of seeing and representing a single mind-independent world, a multinaturalist metaphysics posits a single way of seeing and representing what amount to multiple different worlds, none of which is in the end any more definitively true or real than any other.

And what ultimately makes possible the production of all these different species worlds is "the specificity of bodies," which shape the different "perspectives" of the different species subjects.

> Animals see in the *same* way as we do *different* things because their bodies are different from ours. I am not referring to physiological differences ... but rather to affects, disposition or capacities which render the body of every species unique: what it eats, how it communicates, where it lives, whether it is gregarious or solitary, and so forth. ... The difference between bodies, however, is only apprehendable from an exterior viewpoint, by an other, since, for itself, every type of being has the same form (the generic form of a human being): bodies are the way in which alterity is apprehended as such. ... [I]f Culture is the Subject's nature, then *Nature is the form of the Other as body*.[6]

Finally, this account of the inner logics and dynamics of multinaturalist worlds helps to enrich our understanding of a well-known anecdote once reported by Lévi-Strauss:

> In the Greater Antilles, some years after the discovery of America, whilst the Spanish were dispatching inquisitional commissions to investigate whether the natives had a soul or not, these very natives were busy drowning the white people they had captured in order to find out, after lengthy observation, whether or not the corpses were subject to putrefaction.[7]

As Viveiros de Castro proceeds to explain, these two radically different "experiments" were effectively designed by their authors for the very same purpose, namely to "prove" whether or not the alien others were in fact human persons like themselves. For the Europeans, the presence of a soul would reveal that the Indians were humans not animals. For the Indians, the presence of a mortal body would reveal that the Europeans were humans not spirits.

> The Europeans never doubted that the Indians had bodies; the Indians never doubted that the Europeans had souls (animals and spirits have

them too). What the Indians wanted to know was whether the bodies of those "souls" were capable of the same affects as their own—whether they had the bodies of humans or the bodies of spirits, non-putrescible and protean. In sum: European ethnocentrism consisted in doubting whether other bodies have the same souls as they themselves; Amerindian ethnocentrism in doubting whether other souls had the same bodies.[8]

To a modern western observer, these ontological commitments of Viveiros de Castro's Amerindians may seem to be somewhat exceptional or anomalous, so "extreme" as to be almost the diametric opposite of our own. But they are very far from being the only such "radical alterities" that one can find in the ethnographic record. To broaden our sense of history's ontological diversity and variability, we should now survey a necessarily limited, but representative sample of other instances, ranging in locale from the Americas to Oceania, each of which in some way challenges or subverts the certainties of our own modern way of being human.

Corporate Subjects and Dividual Selves

Consider first, for example, Marshall Sahlins's account of a precolonial Polynesian ontology, where the divine king or chief of the Hawai'ians "encompasses the people in his own person, as a projection of his own being." As the instantiation of the god Lono, his life-giving powers allow him to "assume" and "live" in his own person "the life of the collectivity." In such "heroic polities," the chief is "the condition of possibility of the community," even as his people are simultaneously "particular instances of the chief's existence." And in so far as the lives of the people are thus "calqued upon the powers-that-be," their "biographies are reckoned in dynastic history."[9]

Within the world of the precolonial Hawai'ians, in other words, there could be no free-standing, mutually exclusive fields of state, society, and economy, because social being was unitary, not differentiated. There could be no discrete public and private spheres, because the social content of social being was located in the shared life of a corporate self, not merely in shared spaces where autonomous, self-interested individuals happened to coexist and interact. For that matter, there could be no individuals in the modern sense, since there could be only one such unitary, individual essence, the person of the king himself, of which all his subjects were merely expressions. And there could be no bounded sacred and secular realms, since divinity conditioned the very existence of this one, all-encompassing individual essence, this human corporate subject.

A similarly vital or ontological interdependence between divinity, ruler, and ruled is also intimated by Clifford Geertz in his well-known study of the precolonial Balinese "theatre state" or *negara*. Within the prevailing cosmology, the king was figured simultaneously as the head of a human community and as the center of its "spiritual dimension."[10] As an "approximation" of divinity itself, he was endowed with the power to "ensure [the realm's] prosperity—the productiveness of its land; the fertility of its women; the health of its inhabitants; its freedom from droughts, earthquakes, floods, weevils, or volcanic eruptions; its social tranquility; and even its . . . physical beauty."[11] And as the visible expression of this life-giving power, *negara* court ritual thus tended to assume a "prodigal exuberance," the abundance itself manifesting at once both the source and the results of this royal power.

> The extravagance of state rituals was not just the measure of the king's divinity . . . it was also the measure of the realm's well-being. More important, it was a demonstration that they were the same thing.[12]

Here too, then, there could be no discrete political, social, and economic fields. There could be no separate spheres of public and private, sacred and secular. And there could be no modern-style autonomous individuals. The Balinese *negara* did not empower one individual to rule over others. Rather, it ultimately constituted ruler and ruled as a single, unitary human essence, a monadic social person, which was itself indistinguishable from the royal person of the king. In such an environment, the supposition that humans are all naturally free, self-actualizing subjects would plainly have been unsustainable, if not entirely unthinkable.

In a somewhat more abstract, more analytical vein, McKim Marriott's "ethnosociological" model of traditional Hindu social being draws our attention to a form of subjectivity that was "dividual" rather than individual. According to this account, the person is not conceptualized as an ontologically primitive individual actor with a stable, unified sense of self, but rather as a fundamentally divisible, permeable being, an emergent coalescence of the coded substances, like blood, cooked food, alcohol, money, words, and knowledge, that are exchanged in and through social relations. In other words, the essences of social being are not pregiven human subjects, but the various resources, material and immaterial, which instantiate the relations from which human subjects are made.[13] Accordingly, when persons circulate and exchange these essential resources, they are understood to be altering their own constitution as persons. Their very personalities are permeable and in more or less continual flux.[14]

By Indian modes of thought, what goes on *between* actors are the same connected processes of mixing and separation that go on *within* actors. Actors' particular natures are thought to be the results as well as the causes of their particular actions (*karma*). Varied codes of action or codes of conduct (*dharma*) are thought to be naturally embodied in actors and otherwise substantialized in the flow of things that pass among actors. Thus the assumptions of the easy, proper separability of action from actor, of code from substance . . . that pervades both western philosophy and western common sense . . . is generally absent.[15]

A similarly fictile, dividual notion of personhood seems to have prevailed among the Mexica, the people of Tenochtitlan, Texcoco, and Tlacopan, whose empire came to dominate most of what is now central Mexico during the final century of the Mesoamerican Postclassic period (ca. 900–1521). According to the common sense knowledge of the time, each human being was a compound of harder materials and three lighter "animic" essences.

The first of these animic essences was the *tonalli*, the particular "irradiation" from the sun's rays that was ritually implanted in each neonate child at the time of its birth, thereafter defining its personal fate and identity. The second was located in the liver, seat of the emotions and passions which governed the person's conduct. The third was located in the heart and believed to be immortal in nature, marking the immanent presence of the patron deity who defined the specific human community to which one belonged.[16] Indeed, all categories of beings and things in the Mexica lifeworld were defined by immortal essences of this nature.

The "hearts," or essences, of earthly things were originally the gods of primeval time. Stories tell that these gods wanted to be adored and that the Divine Couple [the Creator gods Ometecuhtli and Omecihuatl] punished their pride by condemning them to inhabit the surface of the earth and the land of the dead. In their new dual home, the condemned gods were subject to the cycle of life and death. On earth these light beings were covered by heavy, hard matter, thus creating humans, animals, plants, minerals, meteors, and the heavenly bodies. As species or classes, they continued to be immortal, but as individuals they went from life to death and from death back to life.[17]

That said, these three components of dividual personhood were not fixed or stable in any ultimate sense. The Mexica person too was permeable, to the extent that a god might choose to invade it, possess it, and alter its constitution for better or worse, variously engendering, say, creativity, lust, illness, or criminality.

Likewise, humans could be ritually transformed into "vessels for divine forces," then sacrificed at feasts as hybrid "god-humans," thereby helping to "sustain the motion of the cosmos." At the same time, the animic essences could leave the body under certain circumstances. The body's distinctive "irradiation" or *tonalli* might escape during times of illness, fright, sleep, or coitus. The ingestion of psychotropic substances might allow one of a sorceror's essences to journey to the heavenly realm of the gods, where past, present, and future were all experienced simultaneously, thus making possible the acquisition of a true knowledge about what was to come. As for death, this was seen as a kind of decomposition of the human person, whereby one's "heart" essence was either abducted permanently by a particular god or departed of its own accord to some otherworldly location.[18] Meanwhile, the material bodies of decedents were merely part of the larger cycle of vegetal life. Eventually they became the maize from which future human bodies would be made.[19]

Dividual forms of personhood have also been commonly identified by ethnographers in Melanesia, most influentially by Marilyn Strathern in *The Gender of the Gift* (1988), her classic study of gender production among the Hagen people of Highland New Guinea and other Melanesians. For the Hagen, as she has shown, all humans are partible persons, each one a living embodiment of the various relations which make their lives possible. As such, each person is an intrinsically social being, the expression of a singular sociality, a kind of ongoing, living product of the sundry actions, gifts, and accomplishments of others that have conditioned their very existence.

> While it will be useful to retain the concept of sociality to refer to the creating and sustaining of relationships, for contextualizing Melanesians' views we shall require a vocabulary that will allow us to talk of sociality in the singular as well as the plural. Far from being regarded as unique entities, Melanesian persons are as dividually as they are individually conceived. They contain a generalized sociality within. Indeed, persons are frequently constructed as the plural and composite site of the relations that produced them. The singular person can be imagined as a social microcosm.[20]

Hence, agency among the Hagen is a matter of personal "decomposition," since the very capacity to act derives from the prior actions of those with whom one has the relations that constitute oneself as a person in the first place. Hence too all Hagen persons can be said to simultaneously contain within themselves essential elements of both genders, since all persons are necessarily products of relations with both males and females.[21] And hence among the Hagen people there is no literal equivalent of our idea of "society." Since each person is an emergent "social microcosm," not a pregiven individual, Hagen sociality is not a function of external relations with a higher order entity, but an internal state or condition, immanent in the very constitution of the subject.[22]

These various ontologies all in different ways challenge our presuppositions about the essential fabrics of social being, especially about the universal self-evidence of human individuals and functionally divisible societal terrains. Others further afield challenge our modern metaphysical commitments in even more fundamental ways. Indeed, Viveiros de Castro's Amazonian multinaturalist ontologies are hardly the only ones that defy modernity's a priori distinctions between the human and the non-human or the cultural and the natural. The Dinka of Southern Sudan, for example, locate the source of all animate life in a generalized "breath," shared by all living creatures, that "comes from and in some way returns to God."[23] Then there are the worlds of hunter-gatherer peoples like the Nayaka of South India, the Batek of Malaysia, and the Mbuti of Zaire, studied by Nurit Bird-David, which are likewise premised upon radically non-modern foundations. In these particular modes of being, which are about as far from our own modern capitalist mode as it is possible to imagine, humans subsist in a "cosmic economy of sharing" with a "giving environment," regarding phenomena likes forests, rocks, and streams as their life-giving mothers and fathers. Apparently, these mostly benign, human-like parents and ancestors will always readily furnish their human "children" with all the means of life, so long as they are honored by feasts, music, and rituals in return.[24]

As Philippe Descola has shown in his recent book, such animist worlds, where humans enjoy "social" relations with all manner of non-human persons or subjects, still to a point subsist in many different global locales, from Amazonia to subarctic Canada and Siberia to southeast Asia and Oceanian islands like New Caledonia.[25] As a final representative example, we might consider the case of the Chewong, who inhabit Pahang Province on the Malay peninsula. Descola describes their "society" as follows:

> Chewong society is not limited to the 260 individuals of which it is composed, for it extends far beyond the ontological frontiers of humanity to encompass a myriad of spirits, plants, animals, and objects that are reputed to possess the same attributes as the Chewong themselves and that the Chewong describe collectively as "our people" (*bi he*). Despite their different appearances, all the entities within this forest cosmos mingle together in an intimate and egalitarian community that, as a whole, stands in opposition to the threatening and incomprehensible world outside, which is inhabited by "different people" (*bi masign*): Malaysians, Chinese,

Westerners, and other aboriginal peoples. Within this saturated intimacy of social life, the beings that share the same immediate environment perceive themselves as complementary and interdependent.[26]

As Descola goes on the explain, for the Chewong, certain animals and plants are "people" (*beri*), sharing with humans a capacity for "reflexive consciousness" (*ruwai*), a kind of subjectivity which "constitutes the true essence of a person and its principle of individuation." As in the case of persons in Viveiros de Castro's Amerindian worlds, the physical form in which each *ruwai* is embodied, whether human or non-human, is in the end nothing more than a dispensable outer "clothing," so that it is therefore possible for the *ruwai* of a human to enter the body of a plant and *vice versa*. Descola duly concludes:

> Not only do distinctions between the natural, the supernatural, and the human have no meaning for the Chewong, but even the possibility of dividing reality into separate stable categories becomes illusory since one can never be sure of the identity of the person, whether human or non-human, that is masked by the "clothing" of another species.[27]

Of course, to all this, one might still be tempted to respond that such radical ontological and metaphysical alterities are surely limited to relatively "simple" or "primitive" lifeworlds. But again this could not be further from the case. In what remains of the chapter, we should consider two further examples, both of them from demonstrably "complex" worlds that are more commonly studied by historians than anthropologists.

The Cosmic Ecology of Ming China

Let us consider first the essential fabrics of the Chinese world that was presupposed and reproduced by the *Great Ming Code* or *Da Ming lü*. Commissioned by the Hongwu emperor Zhu Yuanzhang (1328–1398), founder of the Ming dynasty (1368–1644), this body of laws was first published in 1367 and ultimately finalized in 1397. The immediate aim of the *Code* was to help reestablish a Han Chinese imperial order after the overthrow of the Mongol Yuan dynasty (1271–1368). It thus drew liberally on an admixture of traditional Daoist, Buddhist, and Confucian "teachings," incorporating logics and principles that had first been formulated a thousand or more years earlier, under the Zhou (ca. 1045–256 BC), Qin (221–206 BC), and Han (206 BC–AD 220) dynasties. Though devised to meet the exigencies of a particular historical moment, the *Code* remained in force throughout the

Ming era and formed the basis of the code used thereafter by the Qing dynasty (1644–1911).

In the ontology presupposed by the *Great Ming Code* the fabric of reality resolved itself into a single, all-encompassing cosmic ecosystem, a kind of ongoing symbiosis between three distinct metaphysical realms or entities: a realm of Heaven (*tian*), populated by superhuman gods, spirits, deceased ancestors, and other "cosmic parents"; a realm of "(everything) under Heaven" (*tianxi*), which contained humans, the "cosmic children," and other terrestrial phenomena; and the person and official representatives of the Ming emperor himself, the divinely appointed "Son of Heaven" (*tianzi*), who was authorized by the gods to mediate between the heavenly and the earthly realms. Here, the earthly realm was imagined as a terrain containing three more or less concentric, ontologically distinct zones: a core "central kingdom" (*zhongguo*), which was wherever traditional Chinese ways prevailed; an imperial periphery where life did not yet fully comply with Chinese norms; and the rest of the world beyond. Accordingly, to realize the Mandate of Heaven (*tianming*) upon which his rule depended, the emperor was obliged to conform the entire empire to a single order or mode of life, an order grounded in "heavenly principle" (*tianli*), and then defend its boundaries against all threats from without.

The *Great Ming Code* was thus in essence a device which was designed to perpetuate harmony between the human and the spiritual realms by aligning the conduct of all imperial subjects with the timeless "way" (*dao*) of the cosmos as a whole. Since superhumans were the ultimate bestowers of all earthly blessings and misfortunes, the *Code* sought to maximize the former and minimize the latter by prescribing suitable punishments and incentives for human behavior. In so doing, it functioned not just as a body of juridical constraints, but as a kind of "moral textbook," one authorized by a divinely appointed ruler who hoped to enlist a time-honored true knowledge of the cosmic order of things to bring about a kind of ethical "transformation" (*jiaohua*) in his subjects. For example, in framing the outlawed "Ten Abominations" (*shie*), the *Code* drew upon the "Way of the sage kings of antiquity," which was itself premised upon the "Three Bonds" (*sangang*: ruler over minister; father over son; husband over wife) and the "Five Constant Virtues" (*wuchang*: benevolence; righteousness; propriety; wisdom; fidelity).[28] And for Zhu Yuanzhang, what gave this ancient "Way" its force was precisely its "identity" with "[the Way] of Heaven."[29]

Thus, the *Great Ming Code* at once presupposed and reproduced what has been called a "correlative" cosmology, whereby the human order was essentially patterned after the order of heaven. In the words of Aihe Wang, such a cosmology is an orderly system of correspondence among various domains of reality in the universe, correlating categories of the human world, such as the human body, behavior, morality, the sociopolitical order, and historical changes, with categories of the cosmos, including time, space, the heavenly bodies, seasonal movement, and natural phenomena.[30]

Which is to say, the *Great Ming Code*, when taken on its own terms, was not the instrument of an all-powerful "oriental despot," an autocratic individual who sought to impose his own personal will and interests on an entire population of subjects. Nor was it a device that was designed to remake the world in any sense anew. On the contrary, it aspired expressly to align the conduct of subjects with an always already existing order, one that had in fact prevailed continually since the very creation of the universe, namely the Heavenly Constitution (*tianxian*) itself. To violate the *Code* was thus to transgress the very founding principles of the cosmos.

In the essential, life-sustaining tasks of realizing the Mandate of Heaven in the human realm, mediating between gods and mortals, and harmonizing the empire perfectly with the cosmos as a whole, officials served in effect as extensions of the emperor's body or self. According to some early Ming texts, they were all but ontologically indistinguishable from the person of the ruler himself. In the language of the time, officials were the wings of the imperial swan, the scales and bristles of the imperial dragon, and the very fabrics of the imperial mansion. Alternatively, and more commonly, they were the limbs of a unitary, all-inclusive imperial body:

> The ruler was the head (*yuanshou*); his officials were the legs and arms (*gonggu*); his surveillance and transmission officials were respectively his ears and eyes (*ermu*) and throats and tongues (*houshe*), corresponding to the law-enforcing stars (*zhifa*) in Heaven; and his guards and soldiers were his talons, teeth, armpits, and elbows (*zhaoya, yezhou*). Together these parts shared one heart and constituted a single governmental body. In order for the body to be healthy, the ruler and officials must live with a single heart and mind (*tongxin yide*). Together they served as a cosmic unit mediating between the spiritual and human realms.[31]

Whatever their particular responsibilities, the duties of officials could ultimately be reduced to a single common formula: the "three recompenses and one sacrifice" (*sanbao yisi*). They should "recompense the ruler who granted them authority and wealth, the parents who had given them life, and the people who had supported them with food and clothing; they should also worship the spirits, the overseers of human affairs." If officials abided by these injunctions and virtuously performed their respective functions, cosmic harmony would be maintained and the gods would happily preserve the conditions of human existence, allowing the people to live in peace and prosperity. But if officials were deficient in conduct or character, cosmic disorder would inevitably ensue, with divine displeasure expressing itself through inauspicious celestial phenomena and natural disasters, like floods, droughts, and earthquakes.[32]

So what were the prevailing presuppositions about personhood, the common sense that made possible the idea of a unitary body of ruler and officials just described? How was the human person objectified as a subject and entity at the time?

First and foremost, one's own being as a person was self-evidently inseparable from that of one's parents. As Yonglin Jiang explains, with suitable quotes from contemporary texts:

> As soon as children are born, they owe their lives to their parents, "whose grace is as vast as the boundless Heaven." Children and parents seemingly bear different bodies, but in essence they are "one person": as "blood relatives" (*tianqin*), they breathe the same breath and share the same pulse, and together, they continue the family line. All members of this family line, from ancestors down to future generations, form one common "cosmic being" that is both symbolic and real. In addition, when children treasure the source of their bodies by repaying parental grace and by being filial, their own children will in turn do the same for them: while they are living, they will be supported, and after they die, they will be remembered and served. When this harmonious relationship is established, morality will be promoted and the social order stabilized, people's livelihoods will be guaranteed and consequently, the government's financial burdens will be reduced.[33]

In other words, the observance of "filial piety" (*xiao*) was not a purely "private" or sentimental matter between family members. It had far-reaching consequences for the well-being of the empire, and thus ultimately for the cosmos as a whole. And this helps to explain why the neglect of filial piety was listed as one of the empire's ten proscribed "Abominations."

More generally, the commitment to a fundamentally relational, non-individualist notion of personhood, whereby one's essential being was most immediately conditioned and defined by one's family lineage or unit, also helps to explain why in early Ming practice non-familial relations were routinely conceptualized in terms of familial ties and obligations like filial piety.

To the emperor, filial piety extended far beyond the parent-child relationship. Rather, it involved a broad spectrum of social relations, including self-respect and self-cultivation; husband-wife and senior-junior family relationships; the friend-friend community relationship; the empirewide ruler-subject relationship; and within an administrative region, the official-commoner relationship. By regulating these social relations, the emperor incorporated fundamental dynastic values into the law and brought them to bear upon the single concept of "filial piety."[34]

And such ideas were ultimately rooted deep in Confucian thought, where familial virtues were the basis of any stable imperial order, since "[l]ove for people outside one's family is looked upon as the extension of the love for members of one's own family."[35] As Confucius himself observed: "Filial piety and brotherly respect (*ti*) is the root of humanity."[36] Which is to say, for the Hongwu emperor, as for the Confucian tradition in general, "society is the family writ large."[37]

More generally still, one can say that the very possibility of a modern-style natural, presocial individual, a generic, pregiven unitary human subject or self, would have been all but meaningless, unimaginable even, in early Ming China. According to prevailing Confucian thought on the subject, a person was an altogether more contingent sum of all the particular interpersonal roles and relations which had made his or her life possible and meaningful in the first place. As Henry Rosemont, Jr. explains:

> If I am the sum of the roles I live, then I am not truly living except when I am active in the company of others. As Confucius himself said, "I cannot herd with the birds and the beasts. If I do not live in the midst of other persons, how can I live?" (*Analects* 18.6). [For] in order to *be* a friend, neighbor, or lover, for example, I must *have* a friend, neighbor, or lover. Other persons are not merely accidental or incidental to my achieving personhood and struggling for goodness, they are essential therefore; indeed, they confer personhood on me, for to the extent that I define myself as a teacher, students are necessary to my life, not incidental to it.[38]

Thus, as Herbert Fingarette has written, for Confucius, "the individual is neither the ultimate unit of true humanity nor the ultimate ground of human worth," since a fully human self can only be produced and cultivated through societal roles and relations with others. To put it another way, "where there are not at least two truly human beings, there is not even one."[39]

In sum, this excursus into thought and practice in early Ming China affords us at least a fleeting glimpse of the kinds of ontological and metaphysical alterities that might be realized in a manifestly complex non-modern world. With its all-encompassing cosmic ecology, its life-sustaining consonances between heavenly and human realms, its divinely appointed human mediator between gods and mortals, its unitary ruling body of emperor and officials, its propensity to extend familial virtues to reinforce all societal bonds, and its irreducibly relational senses of personhood and selfhood, the world of the *Great Ming Code* sustained a way of being human that was quite profoundly different from our own.

What then of the complex lifeworlds of the non-modern "West"? Did the peoples of pre-Enlightenment Europe likewise experience life in metaphysical conjunctures radically different from our own? In the third and final part of this book, we shall revisit the case of classical Athens and find there overwhelming evidence for yet another distinctly non-modern way of being human. In the meantime, we can conclude the current chapter by considering a different example, looking briefly at some of the characteristic ontological and metaphysical commitments of western Europeans during the later Middle Ages.

The Medieval Corpus Mysticum

Rather like their counterparts in precolonial Hawai'i and Bali, the inhabitants of later medieval communities understood themselves not as free-standing, presocial atoms but as integral components of larger, continuously existing corporate entities. At least according to contemporary texts, they might know themselves as members of an extended family or, more often, as the members of various social bodies, which could range in size from small, local associations to the unitary body politic of the commonwealth itself.

One of the earliest and best known expressions of this latter idea comes in John of Salisbury's *Policraticus: Of the Frivolities of Courtiers and the Footprints of Philosophers* (1159), where one finds the following, rather vivid figure:

> In the commonwealth, the prince takes the place of the head, subject to God alone and to those who act as His representatives on earth, even as in the human body the head is animated and ruled by the soul. The senate corresponds to the heart, from which proceed the beginnings of good and evil deeds. The offices of eyes, ears, and tongue are claimed by the judges and governors of the province. Officials and soldiers correspond to the hands. Those who are always about the prince are likened to the sides. Treasurers and wardens . . . are like the belly and the intestines, which, if they become congested with excessive greed and too tenaciously keep what they collect, generate innumerable incurable diseases, so that ruin threatens the whole body when they are defective. Tillers of the soil

correspond to the feet, which particularly need the providence of the head because they stumble against many obstacles when they walk upon the ground doing bodily service; and they have a special right to the protection of clothing, since they must raise, sustain, and carry forward the weight of the whole body . . . .[40]

Nor does one have to look far to find similar ways of objectifying the polity in the works of other medieval intellectual luminaries. Evidently part of the common sense of the age, the "body politic" figure is variously elaborated by the likes of Aquinas, Dante, Nicholas of Cusa, William of Ockham, Marsilius of Padua, Christine de Pizan, John of Paris, Jean Gerson, and Sir John Fortescue, to name only a few.[41] Indeed, to a point, all forms of human association could be understood in such terms, from the individual town or city all the way up to the Church and the totality of mankind itself. As Otto Gierke summarizes in his classic study of "The Medieval Doctrine of State and Corporation":

Medieval Thought proceeded from the idea of a single Whole. Therefore an organic construction of Human Society was as familiar to it as a mechanical and atomistic construction was originally alien. Under the influence of biblical allegories and the models set by Greek and Roman writers, the comparison of Mankind at large and every smaller group to an animate body was universally adopted and pressed.[42]

The great English legal historian, F. W. Maitland, explains further:

[M]edieval thought conceived the nation as a community and pictured it as a body of which the king was the head. It resembled those smaller bodies which it comprised and of which it was in some sort composed. What we should regard as the contrast between State and Corporation was hardly visible. The "commune of the realm" differed rather in size and power than in essence from the commune of a county or the commune of a borough. And as the *comitatus* or county took visible form in the *comitatus* or county court, so the realm took visible form in a parliament. "Every one," said [Chief Justice] Thorpe in 1365, "is bound to know at once what is done in Parliament, for Parliament represents the whole body of the realm."[43]

And the very ubiquity and proverbial character of the social body figure in later medieval thought reveals something of the underlying metaphysical commitments which prevailed at the time.

First and foremost, it was taken for granted that human communities were natural formations, not aggregates of pregiven, natural individuals. Since one's very existence and well-being self-evidently depended on other humans, one could not but live in groups. In the words of Aquinas, echoing Aristotle, "because man is by nature a social animal, needing many things to live that he cannot get for himself if alone, it follows that man naturally is part of a multitude, which furnishes him help to live well." As an integral component of a "domestic multitude," he is able to secure life's basic necessities. And as an integral component of a "civic multitude," with its sundry "crafts" and "public authority," he may "not only live but live well."[44] In other words, in this particular metaphysical conjuncture, humans did not confront the world as unitary, self-actualizing beings, free in principle to command and shape their desires, fortunes, and identities as they saw fit. Instead, they were always already defined as parts or expressions of some greater corporate enterprise, conditioned from birth by God, nature, and circumstance to serve the social bodies upon which the lives of all depended, their own included.

Moreover, if the body politic of the commonwealth itself was to live and flourish as such, its members could not be ontologically generic, interchangeable beings. Like the constituents of traditional Hindu castes, all were naturally conditioned for life in a particular "rank" or "estate." As passages like that from the *Policraticus* amply illustrate, all members had to perform their designated functions if all life's needs were to be secured for the community as a whole, from the "prince" as "head" to all the officials spiritual and temporal, right down to the "feet," the "tillers of soil." Thus, the unity and integrity of the polity did not arise spontaneously from some individualist sense of common equality or similitude of condition, still less from any legalistic notion of citizenship or universal "human rights." Rather it was the duty of the head, the ruler, to "dispose" the body politic in the most appropriate manner, to fashion the sundry estates into a single, peaceful and harmonious "order," sometimes called a *corpus mysticum.*[45]

And given these presuppositions, the most "natural" form of rule was plainly monarchy. Again, as in Bali and Hawai'i, monarchy was not objectified as an external domination over fellow individuals, but as the guiding, unifying principle within an extended corporate self or subject. The logic here is succinctly expressed by the exiled Lancastrian jurist, Sir John Fortescue, in *De Laudibus Legum Angliae* (ca. 1470):

As . . . the physical body grows out of the embryo, regulated by one head, so the kingdom issues from the people, and exists as a body mystical, governed by one man as head. And just as in the body natural, as [Aristotle] said, the heart is the source of life, having in itself the blood that it transmits to all members thereof, whereby they are quickened and live, so in the body politic the will of that people is the source of life, having in it the blood, namely, political forethought for the interest of the people, that it transmits to the head and all the members of the body, by which the body is maintained and quickened.[46]

Subsequently, with the emergence of a more centralized, more activist and interventionist form of government under the Tudors, a significant shift occurred in the essential nature of the *corpus mysticum*, when official sources began to claim that the body politic was in fact nothing other than body of the monarch himself. And with this claim was born the curious notion of the "King's Two Bodies," whereby the person of the monarch was deemed to instantiate simultaneously both his own mortal, individual personality and the deathless personality of the realm as a whole, which was "utterly devoid of Infancy and old Age, and other natural Defects and Imbecilities."[47] Thus, during his speech to his first Parliament in 1603, James I felt comfortable characterizing his kingship in the following, rather extravagant terms:

> "What God hath conjoined then, let no man separate." I am the husband, and all the whole island is my lawful wife; I am the head, and it is my body; I am the shepherd, and it is my flock.[48]

Ultimately, a civil war would be fought over the English monarch's claim to embody the entire body politic. And it was precisely this more personalized, absolutist mutation of the medieval *corpus mysticum* which John Locke and other pioneers of classical liberalism would systematically dismember when they later set out to recut the fabrics of western social being. The net result of their efforts would be a radically different vision of the givens of existence, an account that would come to be presupposed and reproduced by our modern capitalist way of being human, by our mainstream natural and social sciences, and thus by the general model of social being that we use to historicize all non-modern lifeworlds.

So given its profound influence upon our practice, let us now turn to examine the ontological common sense of our own modern lifeworld, subjecting its formation and its contents to closer critical scrutiny. We can begin by considering the seismic metaphysical shifts which conditioned its possibility in the first place.

7

The Anomalous Foundations
of Modern Being

The "Great Divide"

When one contemplates the passage from medieval to modern in western European experience, it is easy to be overwhelmed by the many far-reaching shifts in thought and practice that were implicated in this larger process. Even if one were just to consider the period 1500–1800, a highly selective list of landmark changes would have to include, for example: the spread of capitalist modes of production, distribution, and exchange; the growth of alternative, non-Catholic Christianities; the formation of the early modern state; the pursuit of global imperialist projects; the colonial subordination of numerous non-European peoples; the Newtonian revolution in science; the invention of classical liberalism; and the Enlightenment-era promotion of distinctly secular forms of humanism. Yet for all their apparent multiplicity, these various developments all at once expressed and helped to realize a more elemental, more tectonic kind of shift, namely a wholesale reordering of the very foundations of western social being. By about 1800, the timeless certainties of life in a primordial, divinely ordered *corpus mysticum* had been largely supplanted by an entirely new way of being human, one that rested upon the historically novel metaphysical commitments of what we now call modernity.

Philippe Descola has aptly termed this world-historical metaphysical rupture the "Great Divide." And in his view, what ultimately differentiates non-modern from modern worlds is the latter's invention of a radical discontinuity in experience between two mutually exclusive realms, a human realm of culture and a non-human realm of nature.[1] Needless to say, such cleavages in the fabrics of being do not occur overnight. Indeed, according to Descola, the specific conditions of possibility for this nature-culture dichotomy had evolved steadily over the course of two millennia.

Greek philosophers, above all Aristotle, were the first to identify a form of "nature" (*phusis*) as a universal object of theoretical concern. But as yet this new object was merely a developmental process or principle, one that was shared by humans and other terrestrial lifeforms and entities, not a distinct, bounded realm of experience in its own right.[2] During the medieval centuries, Christianity reimagined this unitary object as "Creation," adding novel elements to the picture in the process: a transcendent divinity who was no longer so immediately immanent in his own earthly works; a humanity which was now raised above all companion creatures by virtue of its unique endowment with a god-like reason and intelligence; and as a result, a new sense among humans of inhabiting a world that existed somehow external to oneself.[3] But it was not until the seventeenth century that we witness the birth of "nature as we know it."[4]

For it was at this point that the Scientific Revolution "legitimated the idea of a mechanical nature in which the behavior of every element can be explained by laws, within a totality seen as the sum of its parts, and the interactions of those elements." And it was only at this point that humans began to imagine themselves as the divinely-appointed custodians of the entire natural order. The invention of this nature is summarized by Descola as follows:

> The emergence of modern cosmology results from a complex process in which many factors are inextricably intermingled: the evolution of an aesthetic sensibility and pictorial techniques, the expanding limits of the world, the progress of mechanical skills and the greater mastery over certain environments that this made possible, the progression from knowledge based on interpretation of similarities to a universal science of order and measure—all these are factors that have rendered possible the construction not only of mathematical physics but also of a natural history and a general grammar. Changes in geometry, optics, taxonomy, and semiology have all emerged out of a reorganization of humanity's relationship with the world and the analytical tools which made this possible, rather than from an accumulation of discoveries and a perfecting of skills.[5]

As for the other half of the nature-culture dichotomy, the story of its invention is in a sense more complex still. Even within anthropology, as Descola shows, the term "culture" has been used to refer to at least two ontologically distinct phenomena, each one associated with different national traditions.

In its original definition, manifest in the work of pioneering Anglo-French anthropologists like E. B. Tylor, the term was little more than a synonym for "civilization," in that it encompassed the "knowledge, belief, art, morals, law, custom, and any other capabilities and habits acquired by man as a member of society."[6] In other words, this culture was a singular net product of a universal humanity, one whose early "evolution" could therefore be traced through the study of "primitive" peoples. And to a point, this more universalist notion of culture, which admitted the possibility of a more "scientific" or "nomothetic" anthropology, would sustain the practice of some of the most influential figures in the discipline thereafter, not least Durkheim, Malinowski, and Lévi-Strauss.[7]

The rival formulation is altogether more pluralist and particularistic, whereby "culture" encompasses all of those characteristic norms, values, and meanings that distinguish one historical people from another. As elaborated by the likes of Clifford Geertz and Marshall Sahlins, this is of course the culture that gives its name to mainstream cultural history. And its lineage is distinctly Germanic, extending back through the likes of Franz Boas and Edward Sapir, Wilhelm Dilthey and Heinrich Rickert, ultimately to proto-nationalist thinkers like Herder and von Humboldt, for whom *Kultur* was the product of the "genius" of a given people, not the "civilization" of any universal humanity.[8]

In short, as Descola notes, culture has come to participate in not just one metaphysical dualism but two. According to the first definition, it denotes a realm of human "civilization" whose binary opposite is a non-human realm of nature. According to the second, it denotes a subjective dimension of human meaning and ideation that is forever bracketed off from an objective dimension of self-evident materiality, thereby allowing anthropologists, cultural historians, and others to draw their now customary analytical distinctions between cultural (emic) and material (etic) realities.[9]

For some, of course, Descola's account of a great metaphysical divide may appear to revive the ghosts of the racism and imperialism which haunted the originary evolutionism of the founders of his discipline. But his project could not be further from an anthropology which would "grad[e] peoples according to their proximity to the modern West."[10] If anything, its ultimate purpose is to subvert the very idea of a cultural "progress," by denaturalizing our "naturalism" and insisting on the radical alterity of the modern.

> [T]here is now more to gain from trying to situate our own exoticism as one particular case within a general grammar of cosmologies rather than continuing to attribute to our own vision of the world the value of a standard by which to judge the manner in which thousands of civilizations have managed to acquire some obscure inkling of that vision.[11]

All that said, it is still possible and perhaps necessary to add a little more color, texture, and detail to the Descola's account of the great metaphysical divide itself. It makes perfect sense that someone who has spent a professional lifetime studying the non-modern animist worlds of Amazonia might come to think of modern worlds as quintessentially "naturalist," thereby defining the latter by the presence of that same nature-culture dichotomy whose absence seems to define the former. But if we view this metaphysical divide from a broader historical perspective, I think we can go somewhat further. In what follows, I will suggest that this rupture is in fact sustained by a number of different, if mutually supportive metaphysical commitments. Four such commitments, in particular, seem to set modern capitalist worlds apart from all others. All four are taken for granted by our principal societal structures, including the apparatuses we use to produce our mainstream scientific knowledge, both natural and social. And when taken together, these four foundations of modern being go a long way towards explaining the specific form and contents of the standard ontological template that we use when we historicize the worlds of the past.

Capitalist Metaphysics

First and foremost, our modern social being is founded upon an a priori commitment to materialism. It takes for granted a Cartesian dualism, whereby a primordial dichotomy is inscribed in experience, forever separating matter from mind, object from subject, being from knowing. It thus presupposes an absolute divide between an external, mind-independent reality and the internal thought we use to represent and make sense of that reality. So according to modernity's reigning dualist standards of truth and realness, the only truly real phenomena are physically self-evident, objectively knowable entities, relations, and processes, phenomena which conform to our materialist "laws" of nature and physics.[12]

Hence, within the parameters of our lifeworld, it is simply axiomatic that health and vitality, wealth and value, life and being itself can all in the end be reduced to objectively knowable materialities. And all of our prevailing political, economic, and social mechanisms are duly designed to operate on this principle. As a result, they allow no role for any of the gods, demons, angels, spirits, or myriad other "supernatural" beings that have variously governed, nurtured, energized, and terrorized all other historical realities for millennia. They have no place for those "magical" vital forces that have animated innumerable non-modern ecologies, like Polynesian *mana*, Hindu *shakti*, and Chinese *chi*. For within the confines of our materialist reality, all of these once life-affecting agencies and powers come to look like nothing more than mere fancies of the pre-scientific mind, the irrational imaginings of less "advanced" peoples.

The second of modernity's four fundamental metaphysical commitments is its unapologetic anthropocentrism. As Descola's account indicates, this commitment ordains that all worldly phenomena necessarily belong to one of two pregiven, mutually exclusive orders or realms of experience. Given the apparently manifest exceptionalism of our species, the first of these is of course an order of the human, where all cultural phenomena are concentrated. Defined as the locus of all the world's reason, agency, subjectivity, and societal complexity, this is then the arena where all the business of what we call history is transacted. The second is a self-evidently non-human order of nature, an extramural domain of pure materiality. As such, it tends to be objectified as an ahistorical, self-reproducing "environment" of mute lifeforms, inert "resources," subject-less "processes," and enclosable "property," as a distinctly subordinate domain that we humans are duly entitled to control, manipulate, and exploit as we will.[13]

In other words, by categorically disaggregating the human from the non-human in this way, our largely unexamined anthropocentrism encourages us to believe that we are the chosen custodians of our world, not just another of its incidental products. It allows us to imagine that our modern capitalist mode of life is a purely man-made contrivance, a more or less free-standing, machine-like system of systems that has all but "tamed" an extra-mural nature to its will. And in so doing, this anthropocentrism distances us perhaps irrevocably from countless non-modern peoples of both the present and the past, from peoples whose modes of being were governed by the unchanging annual rhythms of the seasons and the heavenly bodies, from peoples whose very subsistence was synchronized with the life cycles of animals and plants, from peoples who knew the lands that nurtured them in some sense as their parents or ancestors, from peoples who took it for granted that innumerable, mysterious non-human agents and subjects were always out there, immanent in earth, sky, rivers, and seas, making all human life possible.[14]

Much the same could be said of our third metaphysical foundation, namely the secularism upon which our modern western social being is staked. By committing to pursue life in a disenchanted, anthropocentrist environment, one that is always already governed by what we have ourselves determined to be "laws" of science, we obviously rule out the possibility of any meaningful, immediate coexistence with autonomous, all-powerful superhuman actors. In so far as our science can recognize the possibility of divinity at all, it would objectify all gods as purely cultural constructs, as artifacts of human "religion," not as independently existing, "magical" agencies in their own right. Which is to say, within the bounds of our modern real world, humans create gods, rather than the other way round.

Accordingly, to ensure that the subjective, unscientific practice of religion does not interfere with the objective, scientific management of modern populations, we consign it to its own highly circumscribed "sacred" field or realm within the human order, thereby ontologically detaching it from secular fields, like the political, the social, and the economic, where the real business of securing our material existence is to be transacted. Of course, this idea of confining divinities within a single, disaggregated realm of human thought and practice may well make sense to those who have come to think of divinity itself as a mere object of faith or belief, like the god of protestant Christianity. But it would have made no sense at all in most non-modern lifeworlds, where divinity was somehow immanent in all life's processes, where life itself would have ceased altogether if the gods who self-evidently controlled it were somehow relieved of their responsibilities. "Religion" is a category that makes sense only in our modern western world, a world that is already secular, a world where gods have been turned from subjects into objects.[15]

Yet it is the fourth of these metaphysical foundations, our lifeworld's a priori commitment to individualism, that would perhaps be most unfathomable to non-modern peoples of the past. We alone in the modern capitalist west, it seems, regard individuality as the true, primordial estate of the human person. We alone believe that humans are always already unitary, integrated selves, all born with a natural, presocial disposition to pursue a rationally calculated self-interest and act competitively upon our natural, presocial rights to life, liberty, and private property. We alone are thus inclined to see forms of sociality, like relations of kinship, class, nationality, and so forth, as somehow contingent, exogenous phenomena, not as essential constituents of our very subjectivity. And we alone believe that social being exists to serve individual being, rather than the other way round. Because according to the sovereign liberal principles by which we choose to live, human persons are always free-standing unitary essences in the first place, "unsocially sociable" creatures who ontologically precede whatever form of community their self-interest prompts them to form at any given time.[16]

Since this metaphysical commitment to individualism has had a profound conditioning influence upon our capitalist way of life, upon mainstream modern social knowledge, and thus upon historicism's standard ontological template, it would seem appropriate at this point to examine its animating logic in a little more detail. To guide us in the task, we can enlist the help of some key "classical" texts in the Anglo-American liberal tradition, where this logic is quite explicitly expounded.

Dismembering the Social Body

The idea of a natural, presocial human individual was not invented outright by Anglo-American liberals. The proposition that all forms of social being were ultimately reducible to self-realizing, psycho-physical individual beings had earlier been intimated in works by the likes of Hugo Grotius (1583–1645) and René Descartes (1596–1650). And the proposition was first fully elaborated as the basis for a viable social order by Thomas Hobbes (1588–1679), most visibly in his *Leviathan* (1651), where he sought to mount a scientifically informed defence of established forms of absolutist rule against contemporary revolutionary alternatives. In the resulting account, the ideal "commonwealth" amounts to a kind of artificial, secular facsimile of a medieval body politic, one built from essentially modern materials on modern metaphysical foundations.

Imagining a presocial state of nature in which materialist, anthropocentrist, secularist, and individualist conditions of existence already prevailed, Hobbes reasoned that human beings would lack any instinctive disposition towards sociality, being driven merely by their own innate appetites, aversions, and competitive urges to survive and prosper as individuals. A "multitude" of such creatures could only hope to escape a continuous "war of every man against every man" if they formed themselves into a "commonwealth" for their own "peace and common defence." And they could do this in turn only if they enacted a kind of covenant, one that required all of them to alienate their own natural powers of self-protection, to pool these powers together to form a single "common power," and then to confer this power upon a single sovereign figure or body who would duly serve as their representative, thereby embodying "a real unity of them all."[17] Thus, in Hobbes' account, a commonwealth or "Leviathan" of individuals was conceivable only as a quasi-legal *persona ficta*, an artificially constituted body politic, one whose unity was realized only through the very act of its "personation" by a unitary ruling agency.[18]

For early liberal theorists, this idea of an artfully contrived, quasi-medieval social body ruled by a more or less absolute monarch was of course anathema. Given much the same materialist, anthropocentrist, secularist, and individualist conditions of existence, they argued that an altogether different kind of order, one premised upon the exigencies of individual being rather than on those of social being, would emerge more or less spontaneously and inevitably if "nature" were simply allowed to take its course. But to make this case, they first had to complicate the Hobbesian account of the human person by endowing it with some sort of predisposition towards sociality.

John Locke (1632–1704), for example, argued that individuals generally abided by a "law of nature" which enjoined each one "as much as he can, to preserve the rest of mankind" and not "harm another in his life, health, liberty, or possessions."[19] For David Hume (1711–1776), the wellsprings of sociality were already located within an essential human nature, where "selfishness" coexisted with a "faculty" of "limited generosity." And when rationally combined and exercised, these twin faculties would predispose us from the start to develop a "common sense of interest" with others and cultivate social relationships for mutual advantage.[20] Similarly, for Thomas Paine (1737–1809), nature had apparently "implanted" in individuals two distinct mechanisms which "impelled" them "into society," namely a "system of social affections" that are "essential to happiness" and an irresistible urge to satisfy "wants" that cannot be supplied without help from others.[21] And according to Adam Smith (1723–1790), humans were naturally endowed with a power of "conscience" that engendered a "love of what is honorable and noble," thereby tempering the baser urgings of excessive "self-love."[22]

But for our purposes, the particulars of these various accounts of social being are unimportant. If classical liberals could not quite agree on the precise mechanisms which might prompt humans to seek the society of others, all unanimously trusted that such mechanisms existed. Even if their convictions about the essential "faculties" of presocial individuals were based largely on observations of the behavior of the highly socialized individuals of their own time and place, they were evidently intent on asserting a case for humanity's "unsocial sociability" one way or another. Even under primordial, stateless conditions, there would be no "war of every man against every man," as Hobbes had formerly supposed. Liberalism's pioneers insisted that nature herself had somehow endowed humankind with all the resources necessary to form what they now called "(civil) societies," essentially anarchic communities of free, presocial individuals, communities that could subsist and prosper without formal agencies of "government." Nor was this claim merely a theoretical proposition. As far as the champions of early liberalism were concerned, such societies had actually been there all along, fully real and fully present in the material world, but hitherto stifled and suppressed by governments of one kind or another.

How, then, did early liberals go about abstracting ontologically distinct orders of "government" and "society" from a unitary early modern "commonwealth"? To determine where one ended and the other began, they employed a rather crude, almost Manichean relational logic, whereby the two realms were deemed to be antithetical and thus mutually exclusive. Hence, if government was all artifice, the seat of rationally devised laws and institutions, society was a natural order, ruled by the innate, the instinctive, and the impulsive. If government was the public realm of power and prohibition, existing to coerce and compel obedience, society was the domain of individual privacy and liberty. If government in a world of natural individuals inevitably meant rule by self-interested others, society meant the liberty to regulate oneself. And what is perhaps most striking in liberal accounts of this society is the extraordinary latitude of its compass, the remarkable range of practices and processes that were allocated to its precincts under the signs of "nature," "liberty," and the "self."

The general pattern is established early on, in the classic account of the state of nature that is found in Locke's *Second Treatise of Government* (1690). Under such presocial, anarchic conditions, according to Locke, the sheer force of "natural law" would foster an environment of "peace, good will, mutual assistance and preservation," through the creation of shared rules and norms.[23] Here, too, the institution of private property would evolve no less spontaneously, as individuals would mingle toil with soil to make land their own. And here, in time, Locke imagined, some would no less inevitably come to possess more property than others, especially when the invention of money and commerce would make it profitable to own surplus land.[24] Only at that point, it seems, when a naturally self-sustaining community of individuals had already formed, complete with market, graduated property regime, and wealth inequalities, would there be any need to contrive some neutral, coercive agency of rule, since "the great and chief end" of government was simply "the preservation of property."[25] And clearly, there would be no need here for any overbearing Hobbesian authority, because for the most part "society" would already be quite adequately governing itself.

In short, in Locke's arguments, we come close to seeing the very invention of "(civil) society" as an autonomous, essentially anarchic communion of free individuals. In one almighty act of cosmic dichotomy, the ruled were now permanently sundered from their rulers, taking with them almost all of the means for reproducing the conditions of their own existence. And however crudely contrived or epistemologically tenuous this dichotomy might have been, it appears to have steadily acquired the truth of its own self-evidence. When later writers contemplated the ever more convoluted complexity of their own immediate historical circumstances, they now began to see clear evidence for the workings of not one order but two.

Thus, in *Wealth of Nations* (1776), Adam Smith could draw the sharpest of distinctions between the respective contributions of government and society to the material prosperity of contemporary England. Even if the "profusion of government" had undoubtedly "retarded" the country's "natural progress towards wealth and improvement," the "annual produce of its land and labour" had nonetheless risen to unprecedented levels, largely because of "the private frugality and good conduct of individuals," who were all naturally driven "to better their own condition."[26] In the opening broadside to *Common Sense* (1776), Thomas Paine drew an even starker contrast between the private order of society and the public order of government:

Some writers have so confounded society with government as to leave little or no distinction between them; whereas they are not only different, but have different origins. Society is produced by our wants and government by our weakness; the former promotes our happiness *positively* by uniting our affections, the latter *negatively* by restraining our vices. The one encourages intercourse, the other creates distinctions. The first is a patron, the last a punisher. Society is in every state a blessing, but government, even its best state, is but a necessary evil.[27]

And less than a century later, the once revolutionary proposition that civil societies were naturally self-generating and self-sustaining formations, existing ontologically prior to any corresponding realm of government, no longer had to be actively defended. It was simply taken for granted. As Herbert Spencer remarked in *The Social Organism* (1860): "[T]hat societies are not artificially put together, is a truth so manifest, that it seems wonderful men should ever have overlooked it."[28]

Finally, to explain more precisely how such societies were able to subsist and flourish as entities in the absence of government, classical liberal theorists tended to objectify them as self-reproducing life systems. Spencer himself compared them to compound, invertebrate organisms like the Portuguese man-o'-war and the Ascidian mollusk.[29] Other accounts were a little less imaginative. Paine, for example, had argued that a straightforward "mutual dependence and reciprocal interest" engendered a "great chain of connection" between "the landholder, the farmer, the manufacturer, the merchant, the tradesman."[30] And if further proof were needed, he pointed to the recent experience of the young "American states," where "order and harmony" somehow prevailed during the time of the Revolutionary War, despite a relative absence of formal governmental mechanisms.[31] The authors of *The Federalist* (1788) were no less impressed by the seemingly natural solidarity of the young American "nation" or "people," whose bonds had been forged through a shared geography, ancestry, language, custom, religion, and battlefield experience. As John Jay put it, Providence herself seemed to have ordained that Americans be "a band of brethren, united to each other by the strongest ties."[32]

But probably the most consequential case for this kind of stateless social naturalism was made by Adam Smith in *Wealth of Nations* (1776), where he tried to explain how the self-interested, micro-level actions of a population of individuals might unintentionally produce socially beneficial, macro-level outcomes, like the unprecedented levels of prosperity enjoyed by his own fellow Britons at the time. Whereas the productive potential of earlier times had been consistently retarded by politically imposed feudal or mercantilist constraints, the "perfect liberty" of his own era had for the first time allowed individuals to be directed by their respective "invisible hands," their natural disposition to "improve" themselves and to measure their self-improvement by the acquisition of material wealth.[33] And so long as government left all economic actors, from wage-laborers to great landowners, free to act rationally on this disposition, he claimed, the inevitable result would be a maximally efficient, mutually advantageous system of production, distribution, and exchange. The self-serving "invisible hands" of an entire population would more or less spontaneously produce a continually self-adjusting equilibrium between supply and demand, thereby benefiting society as a whole.

That said, Smith and his contemporaries as yet knew nothing of that entity we call the "economy." In their eyes, the processes of production, distribution, and exchange still formed the internal metabolism of society itself. Nature's means of allocating wealth was not yet abstracted from the rest of social life and studied as a homeostatic system or discrete realm of experience in its own right. Indeed, it was not until around the 1870s that social scientists first began to isolate a specialist field of study called "economics," adapting models and categories from physics to explain the dynamics of a "free market" in quasi-mechanical terms.[34] And the idea that there had actually been a free-standing, machine-like "economy" out there all along, a scientifically knowable, measurable, and manageable system of resource allocation, became current only as recently as the 1930s, when it was essentially invented by early econometricians.[35] Nonetheless, the essential conditions of possibility for this "discovery" had been laid some two centuries earlier, when liberal political economists like Smith first set out to explain how a nation of free, naturally appetitive individuals might unwittingly generate unprecedented levels of aggregate wealth.

Such exercises in social naturalism in turn help to explain classical liberalism's peculiar ambivalence towards government. While some form of coercive ruling agency may have been necessary to safeguard and enforce rights, especially the right to accumulate property, it was also by definition a "necessary evil," since it entailed alienating one's natural freedom to rule oneself to other self-interested individuals, who would inevitably rule for themselves.[36] Hence government's powers had to be expressly constrained and rendered accountable by mechanisms like elections, term limits, and rights of recall. Moreover, because the wealth of nations ultimately depended on the unencumbered liberty of individuals to act on their innate dispositions to improve themselves, a free realm of "private" life had to be protected from the realm of "public" power by devices like a bill of rights. And thus in our modern world alone it seems entirely natural that rulers should not manage the basic means of existence, that what we call the "market forces" of our economy should be free from any continuous governmental control.[37]

Modernizing the Fabrics of Non-Modern Being

From this cursory survey of modernity's most definitive ontological and meta-physical commitments, we can now hopefully see more clearly why historicism's standard template of social being assumes the specific form that it does. We can see the historically anomalous logics behind those multifarious divisions which seem to be eternally inscribed on the terrain of our experience, separating the material from the cultural, the human from the non-human, the sacred from the secular, the public from the private, the political from the social, and the state from the economy. We can also see why the template objectifies social being in spatial terms in the first place. After all, in a world of primordial individuated subjects, where sociality is something ontologically posterior to subjectivity, something always external to oneself, it is natural to think of the common spaces where individuals interact, share experiences, and form relations as the ultimate sources or conditions of social being. And we can see why this model seems so intuitively credible and compelling to ourselves as historians. For it reflects the common sense truths of our own everyday lived experiences as subjects, in that it is mounted upon the very same materialist, anthropocentrist, secularist, and in-dividualist foundations that are at once presupposed and reproduced by our own modern capitalist way of being human.

Accordingly, as a device for making sense of social being in our capitalist modernity, this model may well be appropriate and effective, since it is itself a product of modernity's anomalous common sense social knowledge. But that very anomalousness seriously limits the model's utility for historians of other times and places. For it cannot make meaningful sense of worlds which were shaped and controlled by phenomena that our materialist truth standards would require us to deem unreal. It cannot make sense of worlds where humans knew nothing of any bounded, extra-mural order of nature, where non-human agencies and forces continually conditioned the very possibility of human existence. It cannot make meaningful sense of worlds where gods and other superhumans were real subjects, perpetually active in the fabrics of immediate experience, not merely passive objects of faith or belief. And it cannot make sense of worlds where per-sons could never have been generic, presocial individuals, because they were al-ways already products of the particular socialities of the particular communities whose lives they existed ultimately to serve. All of which is to say, this standard analytical device will always be resisted by non-modern ways of being human.

More serious still, we can now better understand the kinds of ontological dis-ruption and damage that can occur when we insist on using this modern template to historicize past experiences. For this ostensibly universal model of social being licenses us in effect to reconfigure the essential fabrics of all those non-modern worlds, to force them all to comply with our own metaphysical commitments to materialism, anthropocentrism, secularism, and individualism.

Thus, when we apply this template, we unavoidably dichotomize the contents of all past experiences along dualist lines, categorically sundering what we deem to be objectively real, material phenomena, like societies, economies, and individuals, from objectively unreal "cultural" or "ideological" phenomena, like gods, myths, mystical life forces, and social bodies. We necessarily dismember the components of non-modern ecologies, detaching what we consider economic, social, and other "human" elements from the "non-human" resources and processes of the "natural environment." We secularize time-honored cosmic hierarchies, rationalizing once-real, omnipotent, divine subjects as unreal, ideational objects, mere effects of the human practice of "religion," which can then be safely confined within its own realm of "the sacred" in an otherwise disenchanted societal terrain. We reengineer the very subjectivities of the human actors on these terrains, stripping them of the non-modern forms of social being that they took to be the essence of who they really were. We then analyze their thoughts, words, and deeds as if they were really natural, presocial individuals all along. And of course these autonomous, quasi-modern subjects will seem entirely at home in the world which our template has made for them, a world that is always already full of quasi-modern social objects.

I therefore submit that it is precisely this modernist model of social being, this most basic tool of our historicist practice, which ultimately causes us to lose non-modern experiences in translation. Far from being an innocent analytical device that allows us to produce impartial, god's-eye accounts of humanity's myriad diverse lifeworlds, this model grants us the god-like power to conduct what amount to elaborate thought experiments upon all past peoples. It entitles us to refabricate all such worlds in the metaphysical image of our own. And needless to say, a historicist practice that requires us to remount the evidence of non-modern experience upon wholly anachronistic modern foundations is methodologically troubling, to say the very least. It is all too obviously liable to produce accounts of past worlds that are improbably riddled with incompletenesses, absences, and unresolved contradictions, much like the consensus account of classical Athens that we examined earlier.

But more fundamentally, this kind of practice is problematic for ethical reasons. Given the past's extraordinary ontological variability, should we not be more troubled by a historicism that is blind to any such variability, that effectively denies non-modern peoples their capacity to determine the real contents of their own experiences? Given the manifest historical anomalousness of the metaphysical conjuncture in which we ourselves have been formed and subsist as beings, who are we to presume that our subjects continually "get stuff wrong," that we today somehow know better? As modernity's designated expert custodians or curators of the human past, should we really be in the business of refashioning experiences in that past to align them with our own modern ontological commitments?

As we saw in an earlier chapter, Descola has described the results of such a practice as "a kind of well-meaning cannibalism," one that "repeatedly incorporate[s] nonmoderns' objectivization of themselves into our own objectivization of ourselves," thereby "wiping out their distinctive contribution to the human condition."[38] I would favor a slightly different metaphor. Far from being an enlightened exercise in preserving objective truths about humanity's many extinct worlds, our conventional practice authorizes us to engage in a kind of retrospective political violence, a historicist imperialism that would forcefully impose the realities of our liberal capitalist present upon peoples who can no longer speak for themselves.

To a point, of course, this kind of imperialism is inevitable in any form of historical practice. After all, even the most ethically sensitive historians must ultimately speak in modern voices, using categories and devices intelligible to modern audiences. But it is one thing to use modern words, as it were, to transcribe the contents of a non-modern reality, trying as far as possible to represent that reality on its own ontological terms. It is quite another to use the categories and other devices of modern social science to translate that same reality into our terms, thereby turning it into something else entirely. Our standard current practice comes at much too high a price for our subjects, effectively denying them their most basic right of self-determination, their ultimate power to determine the essential truths of their own existence.

To this compelling ethical case for an ontological turn, one might of course respond that our historical subjects should have no ultimate power to determine what counts as a world. In the end, their ontologies are merely "cosmological" or "mythological" constructs, the imaginative figments of relatively unenlightened, pre-scientific minds. As historically anomalous as our own modern metaphysical commitments may be, they nonetheless enjoy the endorsement of our mainstream modern sciences, both natural and social. So a historical practice that is premised upon these foundations will still produce accounts that abide by suitably objective, universal standards of truth and realness, however far these stories might diverge from those told by our subjects. As a result, we moderns should remain the ultimate arbiters of all historical truth. Since we alone possess the capacity to frame a truly scientific, synoptic vision of all human experiences across time and space, we will always know better than our subjects what was and what could be really there in their worlds at the time.

Yet as reassuring as such claims may seem, they disregard a rather less comforting fact. For more than a century now, the metaphysical foundations of our modernity and its mainstream sciences have themselves been repeatedly and systematically challenged by numerous distinguished authorities. On closer inspection, these foundations turn out to be no less historically contingent and no more universally true or real than those of any other human lifeworld. Accordingly, as I will try to show in the following chapter, there is also a compelling philosophical case to be made for an ontological turn in historical practice.

8

Ethnographies of the Present

LET US BEGIN with a straightforward question. Where exactly might one find some independent confirmation for our modern western metaphysical certainties, the essential truths which anchor our historicism, our conventional scientific practices, and indeed the whole edifice of our modernity?

Obviously, this confirmation is not to be found in the evidence of our own everyday experience, since that experience is itself already conditioned by precisely those same modern western truths. A phenomenon like our essential individuality as subjects will inevitably seem to possess the truth of its own material self-evidence in our real world, in a liberal capitalist environment filled with phenomena like rights, privacy, democracy, and a free market economy, which all continually presuppose and reproduce the truth of our being as natural psychophysical individuals. Nor can this confirmation come from our mainstream modern science, since that science also already takes the truth of our prevailing metaphysics entirely for granted. A knowledge-producing apparatus that draws categorical distinctions between natural and human sciences, and one that divides human experience into distinct political, social, economic, and psychological fields, has already made its own core metaphysical and ontological commitments all too clear. Nor, for that matter, can confirmation come from any appeal to some transcendent, objective truth standard, since such a standard likewise only has purchase and meaning within a modern dualist metaphysics, one that is already inscribed with that Cartesian line in the sand which would forever distinguish thought from matter in the first place.

If one then looks beyond the confines of orthodox modernist science, to the thought of those who do not take modernity's metaphysical common sense for granted, the picture becomes still less reassuring. Indeed, one will find a disconcerting number of authorities across many disciplines who have all somehow challenged the universal truth status of modernity's prevailing account of what is and can be really there, always already there, in the world. Arguably the most far-reaching of these challenges have come from the field of theoretical physics.

Dualism in Question

Up until the early twentieth century, research in theoretical physics was dependent upon the "classical" mechanistic paradigm of nature, the shared, common sense presumption of an essentially "clockwork universe." The paradigm is succinctly summarized by Shimon Malin:

> The real entities in the universe are particles and fields, existing and changing according to Newton's laws of mechanics and Maxwell's laws of electromagnetism. To explain a phenomenon such as free fall, the tides, or the magnetism of the earth, we have to show how it arises as a configuration of particles and fields subject to these laws. In principle all phenomena are explainable within this framework; hence chemistry is reducible to physics, biology to chemistry, and psychology to biology. And consciousness, which is hardly significant in the immense scheme of the largely inanimate universe, is but one specific item of study within psychology.[1]

The formulation of this paradigm was in turn made possible by a set of universally accepted epistemological and ontological assumptions, whereby all real entities were taken to be empirically observable and measurable things-in-themselves, self-realizing objects that existed independently of any attempts to observe and measure them by human subjects. As the feminist philosopher-physicist Karen Barad explains:

> It is assumed that objects and observers occupy physically and conceptually separable positions. Objects are assumed to possess individually determinate attributes, and it is the job of the scientist to cleverly discern these inherent characteristics by obtaining the values of the corresponding observer-independent variables through some benignly invasive measurement procedure. The reproducibility of measured values under the methodology of controlled experimentation is used to support the objectivist claim that what has been obtained is a representation of intrinsic properties that characterize the objects of an observation-independent reality.[2]

And it is precisely this assumption of a self-realizing, observation-independent reality, the assumption which underwrote the whole enterprise of classical

Newtonian physical science, that came to be seriously challenged in the early twentieth century by the pioneers of a new quantum physics.

To date, probably the most serious sustained challenge has been presented by the so-called Copenhagen Interpretation of quantum physics, which emerged in the later 1920s from propositions formulated by Niels Bohr and Werner Heisenberg and remains the dominant paradigm to this day. This interpretation arose from attempts to resolve apparent contradictions in the essential nature of quantum-level phenomena. Light, for example, will indeed reveal itself as particles under certain experimental conditions, just as Newton had originally claimed in his *Opticks* (1704). But as others like Christiaan Huygens (1629–1695) and Thomas Young (1773–1829) had also shown, when light is examined under different experimental conditions, it appears to take on the character of waves. Likewise, it is impossible to devise a single experimental apparatus which can simultaneously measure both the position and the momentum of a particle like an electron. The more precisely an apparatus fixes the particle's position, the less precisely it can measure the momentum, and vice versa. Bohr and Heisenberg devised different ways to resolve such apparent contradictions. By far the better known of the two is Heisenberg's "uncertainty principle," which frames the issue in terms of the limitations of human knowledge. Essentially, the claim goes, the more precisely we measure one property of a given sub-atomic particle the more we will disturb its other properties, which are all rendered less precisely measurable and ascertainable as a result. But Bohr's alternative, the principle of "complementarity," was even more far-reaching in its implications. By challenging the assumption that sub-atomic particles in fact possess any such essential properties in the first place, properties that are already there to be disturbed, it invites us to rethink the very nature of reality itself.[3]

For Bohr, experiments like those described attest less to the limits of our knowledge than to "the impossibility of [drawing] any sharp separation between the behavior of atomic objects and the interaction with the measuring instruments which serve to define the conditions under which the phenomena appear."[4] In other words, light waves or light particles do not exist as such "in nature," as discrete, objectively observable things-in-themselves. They are rather complementary, mutually exclusive quantum effects. In either case, all that can be said to exist in any objective sense is the "whole phenomenon," the inextricable entanglement of an observed object with an apparatus of observation that produces one or other of these observed effects. And of course the analysis of such whole phenomena requires an entirely new kind of scientific procedure, as Bohr duly observes:

[T]he fundamental difference with respect to the analysis of phenomena in classical and quantum physics is that in the former the interaction between the objects and the measuring instruments may be neglected or compensated for, while in the latter this interaction forms an integral part of the phenomena. The essential wholeness of a proper quantum phenomenon finds indeed logical expression in the circumstance that any attempt at its well-defined subdivision would require a change in the experimental arrangement incompatible with the appearance of the phenomenon itself.[5]

But the momentous philosophical implications of Bohr's complementarity framework are what more concern us here. For as Barad makes clear, it expressly destabilizes our sense of a pregiven, Cartesian divide in experience between subject and object, between observer and observed, the divide upon which modernity's reigning objective truth standard is premised:

What [Bohr] is doing is calling into question an entire tradition in the history of Western metaphysics: the belief that the world is populated with individual things with their own individual sets of determinate properties. The lesson that Bohr takes from quantum physics is very deep and profound: there aren't little things wandering aimlessly in the void that possess the complete set of properties that Newtonian physics assumes (e.g., position and momentum); rather, there is something fundamental about the nature of measurement interactions such that, given a particular measuring apparatus, certain properties *become determinate*, while others are specifically excluded. Which properties become determinate is not governed by the desires or will of the experimenter but rather by the specificity of the experimental apparatus.[6]

In sum, if one follows Bohr's framework to its logical conclusion, one can no longer draw any absolute, categorical distinction between what we know of the world and our means of knowing it. Under Bohrian conditions, objective accounts of "reality" would no longer involve positivist analyses of various mind-independent, materially self-evident entities or essences. Such accounts would instead be no more than unambiguous descriptions of the effects produced by whole phenomena, by interactions between whatever it is that we are observing and whatever apparatuses, concepts, words, and other tools we use to produce and make sense of those observations. Thus, as Bohr himself is said to have remarked:

It is wrong to think that the task of physics is to find out how nature is. Physics concerns what we can say about nature.[7]

Why is it then, one might then ask, that Bohr's propositions and the sub-sequent elaborations of the Copenhagen Interpretation by others have yet to precipitate a general crisis of confidence in the very dualist foundations of our western modernity? Why are our lives conditioned and structured by a mode of social being that still broadly presupposes a "clockwork universe," one that is ultimately reducible to a substrate of material phenomena which all obediently conform to classical, Newtonian laws of physics?

The most immediate answer to this question is surely that classical physics has proved itself to be extraordinarily effective and productive in practice. However questionable its epistemological commitments, it has clearly helped to make possible all kinds of world-changing innovations in communications, wealth creation, data computation, medicine, and so forth. And it is presumably for this reason that mainstream physicists seem largely content to retain classical forms of analysis when examining macro-physical phenomena, apparently believing that quantum considerations apply only at the micro-physical or sub-atomic level. But there is no compelling reason to hold any such belief. As Barad again explains:

> [N]o evidence exists to support the belief that the physical world is divided into two separate domains, each with its own set of physical laws: a microscopic domain governed by the laws of quantum physics, and a macroscopic domain governed by the laws of Newtonian physics. Indeed, quantum mechanics is the most successful and accurate theory in the history of physics, accounting for phenomena over a range of twenty-five orders of magnitude, from the smallest particles of matter to large-scale objects. Quantum physics does not merely supplement Newtonian physics—it supersedes it. The key point is this: Bohr's analysis of the nature of measurement interactions and the epistemological implications of his analysis are completely general (as far as we know).[8]

If so, this generality would then seem to raise the rather vertigo-inducing possibility of a post-Cartesian, quantum-style analytical alternative to "classical" social science, an explicitly non-dualist mode of analysis that would encourage us to see social objects, from modern states and economies all the way down to the modern individual itself, as complex effects produced by interactions between observed materialities and modernity's peculiar ways of observing.

While the possibility of a quantum social science has yet to be widely pursued in any formal way, one could argue that this general analytical direction has already been intimated and perhaps to a point anticipated in the work of a good many theorists for a century or more,[9] To support this claim, we can now briefly consider how some of the most influential critical thinkers of our age have variously challenged the philosophical foundations of mainstream social science.

Critiques of Modern Social Knowledge

Among the most provocative and powerful of these challenges would be those advanced respectively by Karl Marx and Michel Foucault. In their very different ways, Marx and Foucault have thoroughly historicized our prevailing social knowledge, showing quite precisely how the truths of liberal social science, above all the foundational truth of the natural, presocial, appetitive individual, are not the self-evident, timeless, a priori certainties that they are generally taken to be. They are rather at most conditional truths, truths that are thinkable and thus meaningful only within a very particular historical conjuncture, namely under the very specific social, economic, and cultural conditions of liberal capitalist modernity itself.

As Marx long ago observed in his Introduction to the *Grundrisse*, the human being is "an animal which can individuate itself only in the midst of society."[10] Which is to say, there is no such being as a natural or presocial individual. The generic, "presocial" individuals of classical liberal political economy are nothing more than theoretical abstractions. They are merely the logically deduced, individual or irreducible components of a pregiven, higher order entity, i.e., a "society" which, by definition, already exists. More precisely, these human atoms are in fact the highly socialized constituents of European capitalist society, the very same historically specific milieu that happened to produce liberalism's classical political economists.

To understand more thoroughly how these human abstractions have nevertheless come to be realized in modern liberal society, one can then turn to the work of Foucault and his many followers, whose ongoing project of writing a "history of the present" is centrally concerned with the genealogy of the modern individual.[11] Taken collectively, ongoing studies of what Foucault termed liberal "governmentality" and "biopolitics" offer a counterpoint to the conventional, classical liberal story that naturally free and rational individuals have produced the modern world by acting upon their congenital dispositions. They show instead in some detail how the modern capitalist world, through its extraordinarily expansive, decentered circuits of power-knowledge, actively produces the kinds of bodies, freedom, rationality, and individuality that "science" has deemed essential to this world's continued vitality. Simply put, they draw attention to the great paradox at the heart of our modernity, namely the vast amount of "government" of all kinds that is in fact required to produce liberalism's free society of natural individuals. In so doing, these studies not only thoroughly historicize the individual subject as a modern artifact, but they thoroughly destabilize our sense of a world structured by a relatively concrete, objectively existing macro-architecture that would forever divide, say, state from society, government from economy, politics from biology, and theory from practice.[12]

Along broadly similar lines, critical theorists like Timothy Mitchell, Bob Jessop, and Wendy Brown, have questioned the material self-evidence of various modern essences, like the nation-state, state, society, and economy. For when seen from a certain heterodox perspective, one that does not take modernity's prevailing social knowledge for granted, these ostensibly free-standing objects turn out to be no more than complex, historically contingent effects, the commonly sensed outcomes of entanglements between particular hegemonic discursive formations and particular material practices.[13] More generally, proponents of Actor Network Theory, following Bruno Latour, would argue that in fact "we have never been modern." For in their continuing efforts to improve the conditions of modern existence, our sciences have devised new and ever more complex network-like collaborations between human, non-human, and hybrid human/non-human entities or "actants," thereby creating novel objects and relations which would seem to contradict or defy the metaphysical foundations upon which modernity itself is premised.[14]

Elsewhere, gender and queer theorists have expressly tried to formulate novel, post-Cartesian modes of apprehending the relations between the body's cultural representation and its materiality. Most notably, Judith Butler's "performative" account of gender aims directly to subvert the possibility of any essentializing dualist distinction between the body's material constitution and its discursive construction.[15] And broadening our horizons further still, pioneering posthumanist theorists like Donna Haraway and Vicki Kirby, along with philosophers of biology like Susan Oyama and Evelyn Fox Keller, have variously sought to problematize the modernist, anthropocentrist presuppositions and practices which would inscribe absolute, ontological divides between nature and culture, genetics and nurture, environment and self.[16]

At the same time, an ever growing number of works have sought to historicize and otherwise problematize the process of modern, scientific knowledge production itself. For decades now, historians, sociologists and philosophers of science, from Gaston Bachelard and Georges Canguilhem to Andrew Pickering and Bruno Latour, have been scrutinizing how particular scientific apparatuses and procedures objectify and interact with the realities they are designed to study, thereby further complicating our sense of the relationship between observing subject and observed object, between means of knowing and thing known.[17] At a more general level, others in "science studies" have shown repeatedly how the ostensibly objective truths produced by scientific research are variously conditioned by the scientist's own tools, models, assumptions, working environments, and funding sources.[18] Meanwhile, cultural historians like Mary Poovey have helped us to see how modern "facts" all have their own distinct histories.[19] Indeed, as Lorraine Daston and Peter Galison have effectively demonstrated, objectivity itself has a very particular history.[20] And this is not even to mention those leading heterodox philosophers, like Paul Feyerabend, Richard Rorty, and Gianni Vattimo, who have on various grounds dismissed the very possibility of any truly objective knowledge altogether.[21]

As a constructive counterpoint to such portents of epistemological nihilism, the distinctive contribution of Karen Barad should also be mentioned at this point. In her ground-breaking recent studies, she proposes nothing less than a new account of the production of reality itself, by bringing a posthumanist elaboration of Bohr's epistemological framework "into conversation" with recent currents in science studies and critical theory. This alternative account she calls "agential realism." In contrast to the standard scientific realist and social constructivist paradigms, both of which presuppose a Cartesian world that is always already divisible into materialities and idealities, agential realism insists that the "basic units of reality," including the human subject, are just emergent effects, mere "phenomena" that are produced by "intra-acting" material and cultural "agencies."[22] So in contrast to, say, Butler's "performativity" paradigm, Barad's agential realism would extend forms of agency to non-human entities. Rather than seeing materiality as something that awaits inscription with meaning by a world-making human subject, it thereby gives us a means of explaining how precisely "matter comes to matter," as an active participant in the determination of what is and is not there in experience.[23] The net result is a "lively new ontology," whereby "the world's radical aliveness comes to light in an entirely nontraditional way."[24]

All of the above attempts to question and/or reconfigure the philosophical foundations of modern western social knowledge are themselves of course western in origin, and as such these exercises in critical heterodoxy are inevitably shaped and conditioned to some extent by the very orthodoxies which they seek to depose. Arguably even more compelling, therefore, is our final body of critical thought, the various forms of "outsider" critique which draw upon alien, nonwestern reserves of social knowledge.

In recent years, a growing number of western students of non-modern modes of life, like Marilyn Strathern, Philippe Descola, David Graeber, Eduardo Kohn, and Henry Rosemont, Jr., have been inclined to turn the anthropological lens upon the thought and practices of our own modernity, critically scrutinizing our lifeworld as if through the eyes of others.[25] But perhaps more compelling still is the kind of critique with which this study began, namely postcolonial theory, which has no need of any western proxy or mouthpiece. While pioneers in this area, like Frantz Fanon and Edward Said, drew the world's attention to the extremely unenlightened treatment and representation of "others" by liberal, post-Enlightenment Europeans, more recent thinkers in this tradition, like Gayatri Spivak, Dipesh Chakrabarty, Gyan Prakash, Talal Asad, and Gurminder Bhambra, have helped us to see how the parochially Eurocentrist truth claims of the Enlightenment itself have helped to sustain and reproduce colonialist prejudices and discriminations up to our own time. Whether their targets are western knowledge practices in general, social scientific models, historicism, or specific categories like "religion," these critics have shown how our intellectually provincial styles of thought and analysis essentially homogenize all human experience, past and present, predisposing us to objectify it exclusively in western terms, thereby rendering non-western ways of being human essentially indescribable and thus invisible.[26]

Toward a Non-Dualist Metaphysics

Whether or not they address us literally as "outsiders," it may be helpful to think of all of the theorists cited in this chapter in some sense as anthropologists of our modernity, as ethnographers of our present. Scrutinizing our modern western experience as if from without, from a kind of epistemological elsewhere, they together present us with a robust, remarkably variegated counter-knowledge of what we take to be the "real world."

Taken together, these ethnographers of the modern are not suggesting that there is some other more truly real world somewhere out there. On the contrary, they are suggesting that realness itself is something that is always simultaneously "out there" and "in here," so that one cannot definitively say where a subjective inside ends and an objective outside begins. Diverse as their concerns may seem, all thus in some way problematize the primordial, Cartesian divide between mind and matter, between knowing and being, subject and object, the epistemological and the ontological, that makes modernity's objective truth standard possible in the first place. And they do this by showing us again and again how the contents of a real world cannot ultimately be disentangled from the contents of the minds of those for whom that world is real, even if those contents happen to include the belief that the two can in fact be separated.

In so doing, these ethnographers of the present directly or indirectly raise the possibility of a non-dualist counter-metaphysics or "meta-metaphysics."[27] They encourage us to look past the seemingly fixed, stable metaphysical armature of our own real world and see the ultimate plasticity and provisionality of its constitution.

For like quantum physicists of human experience, they help us to apprehend this constitution in terms of historically contingent "whole phenomena," rather than as an aggregation of timeless pregiven essences. Directly or indirectly, they help us to sense the complex ways in which modernity's dynamic observed materialities (practices, institutions, bodies, etc.) combine as reagents with our historically specific "apparatuses of observation" (truth regimes, bodies of social knowledge, subjectivities, etc.) to produce quantum-style materio-cultural effects, like nature and culture, states and societies, economies and individuals, effects that appear to subsist as mind-independent things-in-themselves. And of course, as we saw in the previous chapter, all such effects will be readily realizable as phenomena in a modern metaphysical environment, where the theories and practices of our capitalist way of life already share mutually reinforcing a priori commitments to materialism, anthropocentrism, secularism, and individualism. In this environment, such effects will steadily acquire the truth of their own ontological self-evidence, as if they had really been there all along.

In other words, the proposed non-dualist meta-metaphysics helps us to recognize that our modern common sense knowledge of the order of things is inextricably woven into the very fabrics of our modern being. It encourages us to see how our mainstream dualist science, natural and social, is actively involved in determining where the ontological cuts in these fabrics should fall, thereby helping to constitute the very world that it purports to be merely describing. It thus suggests that this same world is not *the* world but just *a* world, a more or less stable, historically contingent metaphysical conjuncture, one that could not exist as such without the peculiarly modern common sense knowledge that helps to produce it, to make it whatever it really is.

If so, we are thereby released from the dualist commitments of this modern world-making common sense, since those commitments would only have force and purchase within the modern metaphysical conjuncture that they themselves have helped to produce and sustain. We are then free to explore how other such metaphysical conjunctures, other such worlds, might have materialized and flourished elsewhere in time and space, each one of them likewise governed by its own local standards of truth and realness. We are free to see how these non-modern worlds have duly engendered their own non-modern materio-cultural effects, conjuring into existence all manner of ontologically self-evident phenomena, like the gods of the Ming and the Mexica, the dividual subjects of the Hagen, the *corpora mystica* of medieval Europe, the non-human persons of indigenous Amazonia, and the powder-power of Cuban *Ifá*. In short, we are free to accept that there can never be an ultimate, universally true answer to that most fundamental of questions: What counts as a world?

The stakes here could hardly be higher. At the very end of this study, we shall briefly consider how the proposed non-dualist meta-metaphysics has the potential to transform endeavors across many different regions of the intellectual landscape, giving us serious cause to rethink how we produce, organize, and even define what counts as knowledge. But more important for our immediate purposes, this meta-metaphysics grants us the necessary philosophical mandate to engineer a paradigm shift in our own disciplinary practice, effectively authorizing us to write histories of lives lived in many different worlds, not just in one.

9

Ontological History

OVER THE LAST three chapters, the book has laid out ethical and philosophical arguments for analyzing non-modern experiences on their own ontological terms, in their own distinct worlds. In so doing, it has fortified the case made by anthropologists for this kind of move, offering a richer, more representative overview of non-modern ontological alterities (Chapter 6), a fuller, more precise picture of our own historically anomalous metaphysical and ontological commitments (Chapter 7), and a meta-metaphysical license to speak meaningfully of many different worlds of human experience (Chapter 8). As a result, we are now free to contemplate a substantially new, non-dualist vision of our discipline's object of study, one where the variabilities in past experiences would go all the way down to the ontological level.

A New Paradigm

When viewed from this new non-dualist perspective, history would no longer look like the coming-of-age of a single, universal humanity in a pregiven materialist, anthropocentrist, secularist, and individualist environment. It would no longer be the life-story of a unitary species-subject wending its way inevitably towards its full self-realization in capitalist modernity. It would instead be the diverse life-stories of multiple humanities, each one pursuing whatever life meant in its own self-sustaining metaphysical conjuncture, in its own particular world. Every one of these historical worlds, our own dualist world included, would be the cumulative sum of all its own locally produced "whole phenomena," of all the effects generated by entanglements between its own observed materialities and its own world-making common sense. To a point, each one would thus possess its own characteristic way of being human, with its own distinct species of materiality and ideality, subjectivity and sociality, vitality and rationality, and so forth.

Each one would realize its own particular ontological template, its own particular account of the essential contents of experience. And none of these historical worlds thus realized, modern or non-modern, would be any more universally true or real than any other.

Clearly, if one were to embrace this radically different perspective on the lived past, it would have consequences for our practice as historians that are of a most far-reaching kind.

Most immediately, it would require us to relinquish the particular ontological template that has been such a central concern so far in this study. Indeed, it would require us to give up the very possibility of using any kind of universal, one-size-fits-all ontological model when we try to historicize past experiences. After all, if reality in every human lifeworld is the cumulative product of entanglements between its common sense social knowledge and its particular materialities, one cannot make meaningful sense of one such world with models built from the social knowledge of another. For whenever we use such a model to translate a non-modern lifeworld, we are necessarily replacing our subjects' standards of truth and realness with our own, thereby profoundly altering what could and could not be really there at the time. We are fundamentally reconditioning that world, effectively modernizing it at the ontological level. Which is to say, we are in fact fashioning an entirely new lifeworld in the metaphysical image of our own, a strange kind of never-never-world, where the data we selectively abstract from non-modern experience entangles with our own "apparatuses of observation" to produce phenomena that are real only to ourselves, like states, economies, and natural individuals. The net result will always be a kind of hypothetical pre-history of our present lifeworld, not a meaningful "history" of any lived past.

More positively, this new non-dualist perspective on the past would in turn invite us to imagine a new and very different mode of historical practice, one that is premised upon a radically different form of historicism. This alternative historicism would require us to take the ontologies of non-modern peoples just as seriously as we take our own, because whatever our non-modern subjects collectively presumed was always already there in their worlds would be an active, constitutive ingredient of whatever *was* really there at the time. In other words, the fundamental power to determine what counts as a world would be theirs, not ours, to claim. Accordingly, for all analytical purposes, producing a meaningful account of any given past world would require us to embrace the alien standards of truth and realness of that world, to try to make sense of it on its own ontological terms, in its own non-modern metaphysical conjuncture. It would require us to take an ontological turn in our practice.

Of course, this turn to ontological history would mean a fundamental change in the rules of historical engagement, a seismic shift in the very philosophical foundations of our discipline. It would be a shift that is in some ways analogous to the one experienced in the discipline of physics during the previous century, a shift that would likewise be driven by radically different, post-Cartesian ways of apprehending the relationships between subject and object, mind and matter, being and knowing. As we saw earlier, the supersession of classical dualist mechanics by a non-dualist quantum mechanics meant that the enterprise of physics was no longer about disclosing an objectively knowable nature, but about determining "what we can say about nature." If we were then to replace our "classical" dualist paradigm of historicism with a non-dualist quantum-style paradigm, as I am now proposing we do, our disciplinary business would be similarly transformed. It would no longer be about heeding the call of a Ranke to recover an objectively knowable past, a past "as it really was." It would ultimately be about recovering whatever past peoples could say about their present.

So what then might the production of an ontological history involve in actual practice? First and foremost, it would require us to adopt a distinctly recursive mode of analysis, whereby the subjects of our study would determine for us the terms on which they themselves are to be studied. In other words, instead of beginning from a position of pre-knowledge, armed with an array of modern models, categories, and other devices which have already determined for us what we might expect to find in any past world, we would begin from a position of relative ignorance. Our most immediate task would then be to establish the contents of the particular "scientific" template, the local common sense model, that was woven into the fabrics of social being in the lifeworld in question. And to do this, we would have to recover as far as possible the specific terms and categories which our subjects themselves used to objectify the givens of their own existence.

Thus, when trying to pursue a recursive analysis of the life-sustaining thought and practices of any given past people, we would begin by asking a very general question: What would the world have to be like for this particular mode of life to have made sense as such at the time? Our response will then be determined by our answers to an array of more specific questions. For example, what were our subjects' most primordial or elemental metaphysical commitments, the commitments that made possible the world as they knew and experienced it? What criteria did they use to determine what phenomena could and could not be really there? Then, within the metaphysical conjuncture thus realized, what specific phenomena were *actually* there, always already there, in their world? What kinds of subjects, objects, relations, and processes comprised the ontological materials from which the more essential fabrics of their world were ultimately constructed? What kinds of agencies and forces, human and non-human, were active in that world? How did all of these different agencies and forces interact and relate to one another? How were distinctions drawn between them? Upon what kinds of conditions did their respective vitalities depend? How did resources circulate among them so as to reproduce those vitalities? And in the end, what did life itself mean in this particular metaphysical conjuncture? What did its determinative processes necessarily involve? Of what did it irreducibly consist?

Second, having thus established a local template of social being for the world in question, we would then proceed to analyze the distinctive mechanisms of its mode of life on their own ontological terms. For once we have established the particular conditions of existence for this way of life, we should then be better positioned to make sense of its particular norms, practices, and institutions, all of which at once presupposed and reproduced that very same local template of social being.

And if we are prepared to embrace this alternative paradigm and pursue an ontological history along these recursive lines, I suggest, it will yield histories of past lifeworlds that improve on those produced by a classical historicist practice in at least three ways. First, these histories will be more ethically defensible, in that they will seek expressly to restore to extinct past peoples the ultimate power to determine the truths of their own experience. Second, they will be altogether more philosophically robust, since they will now be securely aligned with some of the most powerful currents in contemporary critical thought. And last but hardly least, they will be more historically meaningful. As we shall shortly see in Part III of the book, where we return to the ancient Greek case study, an ontological turn allows us to produce an account of Athenian *demokratia* that is distinctly less problematic and more compelling than the consensus "democratic Athens," not least because it corresponds far more closely to the evidence of experience as it was actually lived and understood at the time. In sum, the new paradigm promises a happy convergence between the exigencies of ethics, philosophy, and history, effectively grounding all three in a single, common principle of human self-determination.

All that said, this proposed shift to ontological history may still beg a few immediate questions. So before demonstrating the move's practical analytical value in the chapters to come, it may be helpful to conclude the second part of the study by addressing some of these questions up front and clearing up any possible misunderstandings.

Q&A

Is ontological history not just a more extreme or uncompromising form of cultural history? What ultimately is the difference between the two?

Essentially, this comes down to a difference between a history of mental representations and a history of experienced realities, between a dualist history of worldviews and non-dualist histories of worlds.

Cultural history remains committed to a conventional modern dualism. It can see only a past of many different worldviews, because it presumes that there is always already a world there to be viewed and represented, a mind-independent material world that is sustained by a more or less constant, transhistorical set of metaphysical conditions. It thus presupposes a historicism that allows for two different kinds of stories about the contents of past experiences, namely etic histories about objective material realities and emic histories about the cultural representations of those same realities. And it thereby keeps intact the presumption that modern historians, with all their modern scientific devices, will always know the true contents of non-modern experiences better than those who actually lived those experiences. It keeps alive the certainty that our subjects continually "get stuff wrong."

By contrast, ontological history would commit us to a non-dualist meta-metaphysics. By embracing a quantum-style account of realness, it would take for granted the active complicity of materialities and idealities in the production of real phenomena, such that the common sense social knowledge of any given past people would then be an essential world-making ingredient of that people's reality, not merely a cultural representation or emic view thereof. It would thus require us to accept that every distinct non-modern people to a point inhabited their own distinct metaphysical conjuncture, their own non-modern world, thereby necessarily restoring to our subjects the final power to determine the truths of their own existence.

In sum, ontological history is not a more extreme or uncompromising variant of mainstream cultural history. Rather, through its commitment to a non-dualist meta-metaphysics, it represents an alternative to an entire disciplinary paradigm, to a dualist historicism that forever obliges us to divide our labors between material history and cultural history in first place. It thus requires us to play the game of history by new rules.

So are you claiming that this new non-dualist, quantum historicism will allow us to recover truly "authentic" accounts of non-modern experiences, as if written from the "inside"?

No, I am not. These accounts will still be histories, products of a peculiarly modern mind and imagination that are expressed in a peculiarly modern voice. They will still in the end be attempts to render into intelligible modern terms whatever it was that past peoples could say about their present. Even the most scrupulously recursive efforts to transcribe the truths of past experiences will still necessarily involve a measure of translation. A historian can never hope to enter fully the subjectivity of someone in an extinct lifeworld, not least because that subjectivity will always have been conditioned by the distinctive metaphysical environment which sustained that particular lifeworld. The best we can hope for is a form of history that, as it were, mechanically reassembles such environments by historicizing past lifeworlds all the way down to the ontological level.

In short, I am not proposing that the very idea of "history" be abandoned altogether as an irredeemably modernist, Eurocentrist enterprise. I am rather proposing an alternative paradigm of historicism, one that is mounted upon new, non-dualist philosophical foundations. And I am suggesting that this alternative historicism will foster a practice that helps us to produce more ethical, more philosophically robust, and more historically meaningful histories, not least because it will expressly deter us from imposing uniquely modern metaphysical conditions upon non-modern worlds.

Yet if we give up the tools of conventional historicism, will this not also mean that we have to relinquish our overall sense of continuities and commensurabilities between different lifeworlds, our sense of a unitary species experience lived in a single spatio-temporal dispensation?

The proposed new paradigm would certainly not rule out analyses of relations between particular worlds adjacent in time and/or space. And it would actively support certain qualified forms of intermundane comparison, especially between the heterogeneous ontological materials from which different real worlds have been made, between the past's many personhoods, human natures, modes of freedom and authority, meanings of life, and so forth. But it would also require us to sacrifice our larger sense of a unitary species experience lived in a single, universal metaphysical conjuncture. It would thus require us to sever the kinds of historical umbilical cords which have hitherto bound, say, the Greeks, the Romans, and other extinct "predecessor" peoples to ourselves. It would require us to see that "Western civilization" is in the end just a story that modern westerners tell about themselves, one that overlooks profound metaphysical discontinuities between worlds. And if these are the prices we have to pay to produce more ethically, philosophically, and historically defensible histories, then they seem to me to be prices well worth paying.

But if our accounts of non-modern worlds are no longer required to abide by modernity's scientific standards of truth and realness, will this not invite a chaos of relativism, a history where "anything goes"?

No it will not. The study contends that all historical worlds, our own included, abide by their own locally contingent regimes of truth and realness, none of which possess any more universal force or purchase than any other. In so doing, it rejects the very possibility of any modern absolute truth standard against which a charge of relativism might conceivably be judged. Moreover, far from portending a history where "anything goes," an ontological turn would require historians to abide by very precise truth standards, albeit those of the particular worlds they are studying, not those of our own western modernity. The project of ontological history is ultimately about retrieving the particular "sciences," the particular bodies of world-making social knowledge, which underwrote the myriad different mechanisms humans have used to produce and reproduce their social being. Again, such bodies of knowledge were woven into the very fabrics of past worlds, not something separable therefrom. Accordingly, as the actual producers of these same bodies of world-making knowledge, our subjects should be the final arbiters of the truth of their experiences, not ourselves.

Nonetheless, if one insists on inscribing a "Great Divide" between modern and non-modern experiences, is there not a danger of reviving the chauvinist, Eurocentrist stereotypes of the Victorian era and re-exoticizing "the other"?

On the contrary, the project of ontological history explicitly recognizes the equal claims to truth and realness of all historically attested lifeworlds, modern and non-modern. It thus implicitly denaturalizes and de-universalizes the metaphysical commitments which sustain modern metanarratives of a common human "progress." Indeed, it raises the possibility of a new kind of critique of our own current conjuncture, one that might allow us to use non-modern, "outsider" criteria to evaluate this anomalous modernity of ours, whether we are considering its signature achievements or the unprecedented levels of devastation that it has visited upon climates, landscapes, non-human beings, human minds, human bodies, and human ecologies through its global experiments in capitalism, imperialism, colonialism, nuclear science, and so forth. In other words, ontological history offers the promise of a critique from a kind of historical elsewhere, one that invites us precisely to acknowledge our own radical alterity, to exoticize ourselves.[1]

Is there not still a more general danger here of essentializing the differences between different lifeworlds?

Assigning general properties or qualities to entire peoples is only problematic in principle if one is already committed to a modern liberal account of personhood, whereby all humans are deemed to be naturally free, self-determining ontological individuals. And when it comes to historical analysis, one could fairly contend that this insistence on a single universal way of being human is in fact a far more troubling form of essentialism. For it works directly against the grain of the experiences of the many non-modern peoples, ancient Greeks included, who took it for granted that their unique pantheons, homelands, bloodlines, ways of life, etc. made them essentially different from others.

Moreover, as a postcolonial critical perspective helps us to see, our modern commitment to an individualist way of being human predisposes us to regard non-individualist modes of social being as somehow under-evolved, less-than-fully-human, and even unnatural, thereby deeply prejudicing our accounts of non-modern lifeworlds.

Realistically speaking, unless one is prepared to commit to some kind of Deleuzian vision of perpetual cosmic flux, where experience is not always already conditioned and anchored by relatively bounded, stable metaphysical foundations, one cannot avoid pursuing a practice that is in some sense essentialist. So the real issue is: Who should determine what the historical essences should be, our subjects or ourselves? For the various ethical, philosophical, and historical reasons already presented, I believe we ought to defer to our subjects on this issue, thereby allowing ourselves to recognize and represent different ways of being human.

But would your insistence on essentializing the differences between human lifeworlds not then impose an unrealistic homogeneity and stability upon the dynamic "mess" of experience within those worlds, forestalling the possibility of any internal diversity or change?

These issues will be more thoroughly addressed in Chapter 16, where they are answered with specific reference to the Athenian case study. For now, the following two points will suffice.

First, the new paradigm would not entail imposing a complete uniformity of thought and perspective upon the inhabitants of each past lifeworld. It would fully allow for a possible diversity of "worldviews" within any given metaphysical conjuncture. But it would insist that the object of the views in question would be the lived world of that particular conjuncture, not what we moderns today take to be "*the* world." It would thus draw a basic analytical distinction between two different modes of ideation. On the one hand, there is taken-for-granted thought about the givens of existence, the common sense that is both presupposed and reproduced by prevailing life-sustaining practices, the social knowledge that is thereby materially implicated in the production of worlds. On the other, there is self-consciously held thought, the beliefs, ideas, and perspectives, orthodox or heterodox, that comprise the particular worldviews of particular subjects. In other words, the new paradigm would firmly distinguish the a priori thought that helps to make a world whatever it is from the a posteriori thought with which one interprets the world thus made. While the former must be relatively homogeneous for a given historical world to subsist, the latter is bound only by the limits of the imaginable in any given time and place.

Second, while the new paradigm would certainly invite synchronic treatments of past worlds, whereby whole periods are viewed as single extended moments in experience rather than as times of eventful sequential change, it would also support analyses of historical discontinuities. Indeed, it would expressly encourage us to apprehend changes in more complex, more multi-dimensional ways, helping us to trace their course on the ontological and metaphysical levels. If, as a non-dualist historicism would suggest, the essential fabrics of experience are ultimately effects produced by ongoing complicities between lively materialities and a people's world-making common sense, then those fabrics will change when such effects begin to dissipate, when a priori knowledge of the world is no longer continually confirmed by the evidence of everyday experience. Likewise, new world-constituting subjects, objects, relations, and processes will come to be realized in experience when new forms of a priori thought become materially implicated in life-sustaining practices, thereby allowing the contents of formerly heterodox worldviews to harden into the contents of worlds.

Thus, for example, when trying to account for that watershed transition from premodern to modern in western Europe, the new paradigm would encourage us to look beyond the standard political, social, economic, cultural, and religious factors. Indeed, it would stir us to identify more precisely the kinds of tectonic changes which made the division of experience into "political," "social," "economic," "cultural," and "religious" realms imaginable in the first place. It would urge us to see processes like the Scientific Revolution, the European conquest of the Americas, the emergence of capitalism, and ultimately the Enlightenment itself as world-making rather than just world-changing developments, as catalysts for novel complicities of knowledge and materiality that would ultimately reorder the essential conditions of existence along new materialist, anthropocentrist, secularist, and individualist lines. And it would help us to see how these new metaphysical conditions duly rendered a vast proliferation of new phenomena suddenly realizable in experience, phenomena that now appeared to have been there, unseen, all along, like orders of nature and culture, realms of state and society, the natural presocial individual, materialist standards of truth and value, and indeed a whole universe of fields, forces, and hitherto invisible particles of matter. In sum, it would help us to document the formation of a historically anomalous metaphysical conjuncture, a world equipped to sustain a radically new way of being human.

While this second part of the study has pressed the ethical and philosophical cases for the proposed ontological turn, the historical case remains somewhat under-developed. Accordingly, the book's third and final part revisits the

Athenian case study to demonstrate how this move would actually work out in practice. Setting aside conventional modern historicist devices, it attempts to show how one can produce more historically meaningful accounts of *demokratia*, the way of life of the classical Athenians, when one views it on its own ontological terms, in its original metaphysical conjuncture.

PART THREE

Life in a Cosmic Ecology

The politeia *is a mode of life (*bios tis*) for the* polis.
ARISTOTLE, *Politics*

10

The Metaphysics of Polis *Community*

DURING THE EARLY months of 403 BC, a growing number of exiles had committed themselves to the cause of restoring *demokratia* in Athens by overthrowing "the Thirty," the Spartan-backed oligarchy which had exercised a tyrannical sway over the *polis* since the end of the Peloponnesian War the previous year. By late spring or early summer, their efforts were starting to gather momentum. Helped by an unseasonal snowstorm, they had successfully resisted attempts by the regime and its Spartan sponsors to dislodge them from their redoubt at Phyle in northwest Attica. Now they found themselves in Mounychia, near Piraeus, in an extremely favorable military position, bracing themselves for an impending attack. According to Xenophon, it was at this point that Thrasybulus of Steiria, the leader of the exiles, rose to address his forces and remind them of both their recent good fortune and the justice of their cause.

> Now we stand face to face with [the Thirty] with arms in our hands. In the past, they had us arrested as we were eating, sleeping, and going about our business in the agora. They had us exiled even when we had done nothing wrong and were not even in the city. And because of this the gods are now clearly fighting on our side (*phaneros hemin summakhousi*). In the middle of fair weather, they sent us a snowstorm, just when it helped us, and just when we attacked, few against many, they granted us the gift of setting up trophies. So too now they have brought us to a position where our enemies, because they are advancing uphill, cannot throw their missiles and javelins over the heads of the front ranks, while we, with spears, javelins, and stones all thrown downhill, cannot miss our targets and are certain to inflict casualties.

And with such help from the gods, we understand, victory was duly won and *demokratia* ultimately restored thereafter.

Such passages remind us of the immense distances which separate classical Athenian experiences from our own. In particular, they help us to recognize that ancient Greek *demokratia*, whatever its precise nature, must have been something categorically different from modern liberal democracy. The distinctions between the two are not just a function of different institutional practices, values, or ideals. Democracy and *demokratia* are products of two fundamentally different metaphysical conjunctures. Evidently, Thrasybulus and his comrades were not protoliberal subjects who were working to restore the rights of natural individuals in a secularist, anthropocentrist world. In their distinctly non-modern environment, *demokratia* was something that could and did depend upon the active support of divinity.

In order to render this Athenian *demokratia* more intelligible, to understand it more on its own ontological terms, we clearly must first try to make sense of the wider metaphysical conjuncture which conditioned its very possibility as a way of life. How then did the Athenians themselves objectify the pregiven fabrics of their lifeworld? What basic metaphysical commitments did they share as a people? What taken-for-granted subjects, objects, relations, and processes was their *polis* made of? According to their prevailing social knowledge, their worldmaking common sense, what kinds of phenomena were really there, always already there, in experience?

One should say up front that answers to such questions are largely to be found by "reading between the lines" of our sources, since antiquity has left us no explicit, comprehensive accounts of what the Athenians took to be the essential givens of their existence. Like other non-modern peoples, the Athenians did not need armies of scientific experts to tell them what was really there and how it was really there. Their prevailing template of social being required no explicit formulation or expression, because it was already inscribed in the fabric of their *politeia*, by which it was both presupposed and duly reproduced. It was implicit not just in the stories that they told about themselves but also in all the shared practices upon which their lives were staked. It was what there was, what could not be otherwise, what did not need to be said. And this template of social being was far less complex and convoluted than our own. As we shall now see, it rested on three primary metaphysical foundations, the first of which was the pantheon of gods in Attica.[1]

Life among Gods

As far as the Athenians were concerned, Attica contained not one community but two. To paraphrase the philosopher Thales, their *polis* was "full of gods" as well as humans.[2] It literally teemed with deities and other numinous beings. Even if not readily visible or even knowable, the gods of the Greeks were not insubstantial principles or ideas. Nor were they merely passive objects of worship and belief. They were real agencies, with their own independent wills and subjectivities, who were everywhere present in the material environment, actively and continually controlling the very conditions of existence. In Athenian experience, "the landscape was a living world, alive with the possibility of divinity."[3]

"[T]he central Greek term, *theous nomizein,* means not 'believe in the gods,' but 'acknowledge' them, that is, pray to them, sacrifice to them, build them temples, make them the object of cult and ritual."[4] Manifestly, immortal gods were always already there, ontologically prior to mortal humans.[5] Since all the wondrous animate and inanimate contents of the sky and the earth, of seas and landscapes, obviously could not have made themselves or been fashioned by human hands, the existence of gods in the world was a self-evident, a priori truth, not a matter of mere faith or belief. If there was a fundamental question for humans to ask, it was not: "Are there gods out there?" It was: "Which particular gods are out there?", and thus "Which gods can we induce to 'hold' our land, to watch over us, and preserve our safety and well-being?"[6] And it seems that the human inhabitants of Attica were unusually adept at enlisting this divine support.

Lest we forget, our proverbially rational, democratic Athenians were no less proverbial among the Greeks of their own time for the number, diversity, and lavishness of their ritual observances, which could involve anything from burning the bones of a hundred or more slaughtered cattle to flinging piglets into potholes, dressing up as bear cubs, casting out scapegoats, and ritually insulting one another.[7] Let us remember too that, in trying to negotiate the challenges of everyday life, these same Athenians were far more likely to have recourse to oracles, amulets, spells, curses, evil eyes, and other such tried-and-true devices than to seek counsel from heterodox intellectuals, from those we would today consider the pioneers of western scientific and philosophical thought.[8] And there were perfectly good, rational reasons for such behavior. The superhuman powers which all these rituals and devices sought to direct to human advantage were not just entirely real to the Athenians. They governed all of life's most critical sources and processes, from conception and birth to health, nutrition, prosperity, and ultimately death itself. Every time one's prayers for, say, a successful childbirth, harvest, or sea voyage were answered, the truth of divinity's immanence in the material order of things was confirmed by its own self-evidence.[9]

To illustrate this immediate material realness of gods in Athenian experience, one could cite any number of ancient texts. To take just one representative, if particularly vivid example, we might consider an inscription on a large stele that commemorated the establishment of the cult of Asclepius in Athens in the later fifth century. As the language of the inscription itself would have it, the monument actually celebrated the "arrival" of Asclepius and his entourage in Athens in the chariot of his human sponsor, one Telemachus, following a journey which had taken them from the god's original "home" in Epidauros via the nearby harbor of Zea in Piraeus. According to the standard reconstruction, the text reads as follows:

> . . . when [Asclepius] came up (*anelthon*) from Zea on the occasion of the Great Mysteries, he was conveyed to the Eleusinion, and having summoned (*metapempsamenos*) his snake from his home, he brought it (*egagen*) hither in Telemachus' chariot. At the same time, (his daughter) Hygieia ("Health") arrived (*elthe*), and in this way the whole sanctuary came to be established during the archonship of Astyphilos of Kydantidai (420/19 BC).[10]

Of course, this "arrival" would have looked very different to a modern observer. And modern historians would likewise be naturally inclined to see here only a succession of human subjects transporting a number of inert, material objects, mere "effigies," "images," or "symbols" of Asclepius and his entourage, by land and sea from one "cult site" to another. But when studying a world which did not subscribe to modernity's materialist standards of truth and realness, we cannot simply dismiss language which explicitly attributes agency and intentionality to gods as just so many "figures of speech." As this text and numerous others indicate, there was no materially self-evident distinction in Athenian experience between the image and the reality of a divinity. Unlike ourselves, the Athenians would not presume to say precisely where the "representation" of a god ended and where the god himself began.[11]

In other words, gods and other superhuman agencies were just as real, just as immediately present, and just as materially efficacious to the Athenians as all of our states, structures, systems, and sundry other man-made, world-ordering phenomena are to us today. The fact that we can never exactly see, say, an object like a state or an economy as such, in all its totality, does not cause us for a moment to doubt its objective, mind-independent existence as a hypostatic thing-in-itself, so long as we believe we can sense its presence and its effects all around us. So it was with the Athenians and their gods. Perhaps the biggest difference was that, under certain special circumstances, the gods of ancient Athens were actually observable.[12]

If there was a characteristically Athenian attitude towards the gods of the Greeks, it was not one of bemused intellectual skepticism towards such naïveties. It was rather the attitude visible in the story broadcast in the west pediment of the Parthenon, the story that two gods had once seen fit to compete to become the divine patron and protector of Athens, with the Athenians themselves ultimately deciding the issue, when they favored Athena over Poseidon and duly took her name as their own. It was the attitude visible in the rather bold, central tableau of the famous Ionic frieze on the same building, which figured all of the Olympian gods dutifully arrayed in a semicircle to watch mortal Athenians as they staged the grand spectacle of the Great Panathenaia, the premier festival of the *polis*.[13] And it was the attitude discernible in the Athenians' general, common sense conviction that they and their land were unusually "god-beloved" (*theophiles*). Indeed, to judge by evidence for ritual activities, there were at least two hundred distinct deities present in Attica during the classical era, more than in any other Greek *polis*.[14]

So if, among Greek peoples, the Athenians may seem unusually prone to self-regard and self-aggrandizement, this was not because they were in any sense "humanists." They were self-regarding precisely because they were convinced they were unusually favored by divinities, not because they presumed they could somehow manage their *polis* without them.

No surprise, then, that the most prestigious, most costly architectural expressions of Athenian reason were not deliberative chambers or philosophical schools. They were temples, appropriately opulent residences for the beings who were expected to direct the lives of all in Attica. And no surprise that most Athenians in any given year expended significantly more time, energy, and resources on engagements with divinity than they did on, say, deliberations in the law courts, council, or assembly meetings.[15] Unlike their counterparts in modern lifeworlds, the Athenians did not presume that humans had been granted the powers necessary to govern the essential vitality of entire populations. In their world, the most urgent work of vitality management was accomplished by ritually inducing others to do it, namely the non-human beings who did possess such powers.

Nor is it any surprise that even the comparatively modest levels of human "governmental" activity that did go on in Athens were always accompanied by oaths and oracles, rites of purification, prayers, sacrifices, and the like, or that the items of business transacted by Athenian decision-making bodies regularly included management of the community's relations with the divine.[16] In such an environment, a separation between "church" and "state" would have been entirely irrational and self-defeating, if not entirely unimaginable. Even if one could somehow exclude the gods from a human "field" of collective decision-making, why on earth would one want to? Without their continuing support, the outcomes produced would have been quite meaningless. And the same might indeed be said of every other communal event and undertaking in Athens which routinely featured ritual actions, including

[t]he testing of qualifications for citizens and officials, the formulae and agenda for meetings of the [assembly], military campaigns and individual

battles, legal procedures and treaty-making, marriages, symposia, festivals, dramatic performances, sporting events, meals, farming, and even trade (through sacrifices or libations before journeys and sacred tolls, harbor dues and the like) . . . .[17]

In short, we see in Athens nothing less than the "total suffusion of . . . society with the gods and their concerns."[18]

All of this is common scholarly knowledge. And as such, it presents signifi-cant challenges for mainstream historians, who, it goes without saying, do not readily accept the reality of Greek divinities.[19] How exactly can one account his-torically for a past world where the primary actors were apparently unreal, ahis-torical beings, pure figments of the imagination?

The standard solution is of course to translate the presence of the divine in Athens into "religion," thereby consigning it to a neatly bounded field of purely human thought and action. With this field thus safely disaggregated from other such fields, like the political and the economic, one is then free to treat the stuff of "history proper," like decision-making, warfare, courtroom disputes, trade, and agriculture, as if they were wholly secular activities taking place in a largely dis-enchanted environment, as if divine conditioning was no part of whatever these phenomena actually were at the time. And if religious activity itself ever threatens to intrude upon the stage of our history proper, one can always follow time-honored precedent and seek ulterior motives behind such conduct, rationalizing it as a form of self-worship, a mechanism of social cohesion, a quest for order amid chaos, and so on. Or else one can simply ignore such activity altogether and leave it to specialists in the subject, who likewise tend to reify "religion" as a free-standing field or system in its own right.[20]

But as we have already seen, there was no real equivalent of the category "re-ligion" in Greek thought or language. Nor was there any obvious need for such a category in a lifeworld that was "suffused with the gods and their concerns," where divinity ultimately controlled the sources, means, and ends of life itself. And as we also saw earlier, the standard alternative position, which is to regard religion as an "embedded" phenomenon, only raises further problems of an onto-logical nature. All of which begs an important question. If we cannot say precisely how religion was "there" in Athens because it was nowhere as such, why do we still insist on using this category? If we must reconfigure the basic parameters of Athenian reality just to make this category somehow fit, what exactly do we gain by applying it at all? What if the real source of our problems here is the category "religion" itself?

This category may make perfectly good sense in our own materialist, anthropocentrist, secularist, and individualist conjuncture, where the very

existence of divinity is purely a matter of private "belief" or "faith," where social being as a whole is thus not staked upon active, perpetual divine support,
where some significant portion of this social being can therefore be objectified as
"not-religion." In such an environment, it is not hard to see how polities might
have little compunction consigning human-divine relations to a relatively limited, clearly delineated space within their societal terrain. Seen from the inside,
through the eyes of "believers," this space may still seem to vouchsafe the truth of
an entire "sacred" reality, one in which a somewhat detached divinity still rules
from afar. But when one views this non-secular field from the secular outside, one
sees only a collection of all-too-human individuals, all alone with their "unscientific" beliefs and practices. In short, one sees only "religion," just another functionally specialized domain of social experience within a much larger otherwise
secular totality.

Indeed, there are compelling reasons to accept recent claims that "religion" as
we know it is an invention of western modernity. For religion only makes sense
as a category in a world that can already imagine containing the divine within
a specially designated field of social life, a world which has reduced godhead to
an effect of the thoughts and actions of human beings. Which is to say, it makes
sense only in a modern world, where the prevailing social knowledge no longer
accepts the self-evident, mind-independent reality of divinity itself, not least because it presumes that humans can secure their own vitality largely if not entirely
by themselves.[21]

Either way, the category "religion" makes no sense at all in a context like
classical Athens. In the world of the Athenians, the realness of divinity was not
a subjective matter of personal faith or belief. It was an objective fact, the most
fundamental, most indisputable fact of their existence. What may look to us
like "religion" was present everywhere in their world because divinity was manifestly present everywhere in their world, immanent in its very fabric, continually
controlling the mysterious wellsprings of life itself. Such a world could not be
conveniently divided into parallel sacred and secular realities, each with its own
distinctive ontological contents. There was only one possible reality, with one
possible ontology, and it was full of gods.

To apply the category "religion" to such a world is thus not simply to misrepresent what was really there. It is to alter what was there. It is to insert into
an ancient reality a modern social object that was not and could not have been
there, an object whose presence profoundly affects how we then think about everything else. For one cannot admit religion to the *polis* without also importing
with it the kinds of metaphysical conditions under which it would make sense
as a phenomenon. And this means imposing upon Athens an essentially secular
ontology, where gods are effectively reduced to inert human constructs, where

communication with divinity is thereby reduced to a marginal, largely self-referential or "symbolic" activity, and where ritual is accordingly divorced from all the other business of real, everyday life, the very life which it existed to help govern.[22] In short, the category "religion" positively hinders our efforts to produce historically meaningful accounts of classical Athens. There are, I submit, the strongest possible reasons for questioning its analytical utility.

So what was really there in Athens, if not religion as such? How and where did the Athenians make the necessary cuts in the fabric of reality to accommodate the presence of gods in their world? Happily for us, the Athenians were very consistent in the way they conceptualized the contents of their *polis*. Time and again they use the very same binary classification to render the entangled complexity of their world intelligible and thereby actionable. Time and again they invoke the very same pair of categories, whether they are specifying what exactly it was that all Athenians shared as Athenians, what it was that outsiders gained when they became Athenians, what it was that Athenians were defending when they defended their *polis*, or what it was that Athenians could potentially lose if they were defeated, governed poorly, or succumbed to the personal rule of a "tyrant."

To take an example almost at random, Demosthenes describes as follows how a law supposedly authored by Draco (ca. 620 BC?) required the physical exclusion of anyone guilty of unintentional homicide from Athens:

> [Draco] keeps the offender away from everything in which the deceased used to share in his lifetime; first, from his fatherland (*tes patridos*) and from all things therein (*ton en tautei panton*), both the *hiera* and the *hosia*, assigning the frontier-market as the boundary from which he declares him excluded; second, from the sanctuaries held by amphictyonies . . . [23]

What, then, were these *hiera* and *hosia* that comprised "all" of the "things" in the *polis*?[24] The conventional answer to this question is that they referred to all things "sacred" (*hiera*) and "secular" (*hosia*). But since, as we have seen, this distinction presupposes a modern western metaphysical conjuncture, the ancient terms must refer to something else. And various recent studies would support this supposition.

When viewed in context, the category *ta hiera* seems to refer to all those subjects, objects, relations, and processes in the *polis* that could be said to belong in some definitive sense to its divine inhabitants. In practice, the term was used to designate all of those things which the human population had consecrated or otherwise set aside for their presiding gods to use, enjoy, or claim as their own, usually through some form of ritual purification. For by definition, *ta hiera* were things that were "pure" or "immaculate" (*kathara*), free from "pollution" (*hagna*), undefiled by death, blood, and other phenomena anathema to immortals. The category thus included, for example, all of the temples, sanctuaries, altars, cults, festivals, sacrifices, votives, treasures, ritual paraphernalia, and sometimes even the persons that were "dedicated in a special way" to the gods.[25]

As such, *ta hiera* were distinguished from things which were somehow innately venerable, numinous, or divine (*ta hagia, ta semna*), things which thus strictly speaking could not be considered part of any human *polis*.[26] But more immediately, within the confines of such a *polis, ta hiera* were also categorically differentiated from *ta hosia*, meaning all things that had not been rendered *hiera* by some form of ritual purification. In other words, by a process of mutual exclusion, if *ta hiera* comprised all of those things which the divine inhabitants of Attica could claim as their own, one might say that *ta hosia* included all of those things which the gods were willing to accept belonged in some sense to the human inhabitants. As many scholars have shown, the adjective *hosios*, when used independently, had a semantic range that seems to encompass ideas like "pious," "lawful," "approved," "correct," and thus liable to elicit approval from the gods.[27] The noun *ta hosia* accordingly encompassed all subjects, objects, relations, and processes in human experience that were not formally "purified" but still somehow conformed to divine prescriptions and expectations. In short, to objectify the totality of a *polis* as *ta hiera kai ta hosia* was to dichotomize its contents into things that were either acceptable to the gods or belonged to them outright.

This binary classification is thus fundamentally different from our modern sacred-secular distinction in at least two ways.

First and foremost, both of the Greek categories presuppose a reality that is ultimately conditioned and ordered by divinity. In such an environment, there can be nothing wholly "secular" or purely "profane." Even those things which humans can be said to possess are theirs to dispose only insofar as they are used in ways and to purposes that are *hosia*, or pleasing to the gods.

Second, and related, the Greek categories do not refer to two entirely separate, mutually exclusive "spheres" or "realms" of social experience. The two were very often involved together in the very same experiences, in ways that were mutually productive, not mutually conflicting. For example, when performing sacrifices or other *hiera*, one would always strive to perform them in ways that were *hosia*, or conducive to divine favor, by following all the correct, prescribed procedures.[28] Likewise, one could freely draw on one's own *hosia* resources for the *hosion* purpose of performing *ta hiera* in the first place.[29] So too one could, say, borrow from the *hiera* in the treasury of Athena herself for the *hosion* purpose of, say, defending her *polis*, the "god-beloved" *polis* of the Athenians, from its enemies.[30] And needless to say, to observe or preserve the very distinction between *ta hiera* and *ta hosia* was itself an act that was *hosion*.[31] Because ultimately these were the terms which specified how gods and humans could exist together in the same world. One might say they specified what belonged to each of these two coexistent communities when seen through the eyes of the other.

So the two categories did not evoke entire parallel realities, organized upon diametrially opposed principles, as the sacred and the secular might do in our world today. The world of Athens was innocent of any such man-made cosmological contradictions. In Athens, *ta hiera* and *ta hosia* were two complementary, constituent ingredients of the same reality, because there was and could be only one single reality, one in which both mortals and immortals subsisted, ideally in a condition of mutual dependence and support. All of which is to say, the Athenians did not objectify their *polis* as an artifact of their own making, as an abstract, rationally ordered "society," as a functionally differentiated formation that was created exclusively by and exclusively for humans. They saw it rather as a single, unitary system of life, a divinely ordered ecology, through which a pantheon of gods sustained a community of humans who pleased them.

So where did the life which flowed through this system ultimately come from? What were its primary sources and means? Whence the very stuff that sustained the vitality of the cosmic ecology of classical Athens? This brings us to the second foundation of Athenian social being, namely the ancestral land of Attica itself.

Mother Attica

To modern eyes, ancient Attica tends to possess the same general, universal properties as any other inhabited earthly tract. Depending on one's interests and perspectives, one might see it as a usable "natural resource," as a body of cultivable land that was capable of sustaining both pastoral and arable forms of production. One might see it in more abstract economistic terms as a commodity, as a tract of enclosable public or private "property," that could be owned, exploited, and exchanged. Or one might see it in more political terms, as the particular "territory" of a particular polity. But in the eyes of the Athenians, Attica was not a particular instance of any supposedly universal phenomenon. It was a land like no other, a singular, living organism that was ecologically and even genetically inseparable from themselves.

Uniquely among mainland Greeks, the Athenians knew themselves as an indigenous or autochthonous people. According to the conventional wisdom of the age, they alone had inhabited their homeland continuously since time immemorial.

> For alone of all mankind they settled the very land from which they were born and handed it down to their descendants. Thus one may reasonably

assume that whereas others, who came as immigrants into their *poleis* and are deemed to be members of the same, are comparable to adopted children; but the Athenians belong to their fatherland (*patridos*) by virtue of legitimate birth.[32]

This passage also reveals the primary figure invoked by the classical Athenians whenever they tried to make sense of their relationship with their land, namely the figure of the family. The funeral oration credited to Lysias is even more explicit:

[At] the beginning of their life (*he arkhe tou biou*) . . . [the Athenians] had not been collected, like most peoples, from every quarter, and had not settled in a foreign land after driving out its people: they were indigenous (*autokhthones*), and possessed in one and the same land their mother (*metera*) and their father(land) (*patrida*).[33]

This familial figure also found more vivid expression in the claim that the first Athenians, notably the primordial, snake-tailed king Cecrops and his successor Erechtheus/Erichthonius, were quite literally *gegeneis*, "born from the earth" of Attica.[34] And the figure could in turn be further elaborated and generalized into a claim that the Athenians as a whole, as the "descendants" or "people" of Erechtheus (*Erekhtheidai*), were themselves ultimately the progeny of their native land.[35]

[W]e are of a lineage so noble (*kalos gegonamen*) and so legitimate (*gnesios*) that throughout our history we have continued in possession of the very same land from which we have sprung (*ex hosper ephumen*), being its indigenous inhabitants (*autokhthones*), and are thus able to address our *polis* by the very names which we apply to our nearest kin; for it is appropriate for us alone of all the Hellenes to call the same *polis*, our *polis*, at once nurse and fatherland and mother (*trophon kai patrida kai metera*).[36]

Some modern observers may be uncomfortable with the thought that the otherwise enlightened Athenians might seriously have entertained such apparently "magical" or "supernatural" origins for themselves.[37] But theirs was a different metaphysical conjuncture, a reality of gods and men, of *hiera* and *hosia*, a reality that was not circumscribed by any secular "laws" of physics or nature. And according to the evidence, the idea of something like a consanguinity between people and homeland became part of Athenian world-making common sense during the classical era.

At the same time, the familial figure also speaks to the Athenians' profoundly non-modern conception of their land as a kind of living person or agency. If the notion of "fatherland" (*patris*) genders the land as an inert inheritance, passed through the male line of a single extended family unit, the claim of Attica's maternity objectifies the land as an active, characteristically female body of generative earth or soil.[38] In the minds of the Athenians, it seems, mother Attica was a vital and constant participant in the life and history of her offspring people, containing within herself a certain subjectivity and remarkable powers and properties.

Most remarkably, she had been the first land to produce crops of grain and olives, and thus the first to sustain human life of any kind.

> [D]uring that period in which the whole earth was putting forth and producing animals of all kinds, wild and tame, our country showed herself unproductive and devoid of wild animals, but chose for herself and gave birth to man, who exceeds all other animals in intelligence and alone of all creatures recognizes justice and the gods. . . . For every creature that produces young possesses a suitable supply of nourishment (*trophen epitedeian*) for its offspring; and by this criterion it is clear also whether a woman be truly a mother or not, if she possesses sources of nourishment (*pegas trophes*) for her child. Now our land, which is also our mother, provides full proof of her having brought forth men; for, of all the lands that then existed, she was the first and the only one to produce human nourishment, namely the grain of wheat and barley, whereby the whole of humankind is most richly and well nourished, inasmuch as she herself was the true mother of this creature. . . . And after it she brought to birth for her progeny the olive, a salve to their pains.[39]

Nor did the wondrous generative powers of mother Attica cease there. The same speech continues:

> And when she had nurtured and reared them up to maturity, she introduced gods to be their rulers and teachers (*arkhontas kai didaskalous*); . . . and they set in order our mode of life (*ton bion hemon*), not only in respect of daily business, by instructing us before all others in skills (*tekhnas*), but also in respect of the defence of our land, by teaching us how to acquire and handle arms. Such being the manner of their birth and of their education, the ancestors of these men framed for themselves and lived under the order of a *polis* (*oikoun politeian kataskeuasamenoi*).[40]

In other words, by ensuring that the Athenians were not only able to feed themselves, but to defend and administer all the contents of their *polis*, mother Attica effectively equipped her people to work for and secure her own continued vitality. Much like their relations with the gods, the Athenians' relations with their maternal soil were thus fundamentally reciprocal. Just as Attica was "the very nurse of their existence," so too they "cherished her as fondly as the best of children cherish their fathers and mothers."[41]

These feelings of consanguinity and reciprocity between people and land were expressed in a range of different Athenian practices. Most immediately, ritual offerings to the divine Mother Attica, as Ge Kourotrophos ("Land, Nurturer of Youth"), were a routine feature of festivals all over Attica.[42] Such feelings also help to explain why the Athenians were so unusually insistent on interring their dead in the earth of their motherland. After spending their lives "cherishing" the very source of their existence, the autochthonous inhabitants of Attica departed those lives by returning to this same soil, the very substance from which they had ultimately been made. An ongoing cycle of life was thereby perpetuated, a process of ecological reproduction which to a point denied the possibility of any absolute material distinction between land and people. Hence too, military commanders were required, on pain of death, to retrieve the bodies of all those who had died fighting for the *polis* abroad for formal burial in Athens.[43] According to the prevailing logic of the annual mass funeral that followed, the war dead were

> nurtured . . . by that motherland (*metros tes khoras*) wherein they dwelt, and now rest in death in the domestic abodes (*en oikeiois topois*) of the mother who bore them, reared them, and welcomed them back.[44]

A sense of the land of Attica as a generative agency, as an active participant in the ongoing life processes of a *polis* ecology, also helps to explain why the bodies of murderers could not be interred in Attic soil. Persons polluted by bloodshed could have no contact with the *hiera* of the gods, so to bury such a person was to risk to incubating the pollution and contaminating the entire land of the *polis*, the very source of its material well-being, thereby potentially rupturing relations with life-sustaining divinity itself.[45] For much the same reason, even an inanimate object, like a falling rooftile, that had somehow caused a human death had to be ritually expelled beyond the physical bounds of the *polis*.[46] And a similar kind of logic likewise forbade the burial in Attica of traitors and temple robbers.[47] As Lycurgus explains to the panel of judges in the opening to his prosecution of one Leocrates for treachery (330 BC), Athenian traitors do not just betray their own land and people; they betray "Athena and those other gods and heroes whose statues are erected in our city and the country," they betray "their temples, shrines and precincts," and they betray "the honors which your laws ordain and the sacrificial rituals which your ancestors have handed down."[48] In sum, to bury the body of such an egregiously impure person in the soil of mother Attica would have been to imperil the very existence of the *polis* as a whole.

Such passages as the last in turn reveal something more of the Athenians' general metaphysical presuppositions, the larger, pregiven truths of social being upon which their whole "way of life" or *politeia* was premised. The picture that begins to emerge is one of a kind of symbiosis between three foundations or essences, a cosmic ecology maintained at once by gods, land, and people. Within this general metaphysical scheme, there could be no absolute divisions between categorically distinct orders of experience, between sacred and secular, natural and cultural, human and divine. There was just one order of experience, a single system of *polis* life in which all three essences were continually and mutually implicated.

If the gods' role in this system was to reproduce the basic conditions of existence for land and people, the role of mother Attica was to furnish the essential means of material life for people and gods. Accordingly, all usable land (*ge*) in the *polis* could be categorized as either *hiera* or *hosia*, depending on which of the two populations was deemed to control its disposition.[49]

The *hiera ge* designated for the gods' use took the form of *temene*, plots that had been "cut off" (*temno*) or circumscribed to serve their special purpose. Such *temene* included all spaces occupied by sanctuaries of the gods, along with the cultivable lands leased to maintain the cults of those sanctuaries.[50] They also included certain special places that could not be cultivated because of their powerful, historic associations with particular divinities, such as the domestic precinct of Athena Polias herself on the acropolis, the grove of the "August Goddesses" (*Semnai Theai*) at Colonus, the "Holy Meadow" (*Hiera Orgas*) west of Eleusis which had (presumably) incubated Demeter's first grain crop, and the plots occupied by all the sacred olives (*moriai*) that were descended from Athena's original arboreal gift.[51] The rest of Attica, meanwhile, could be considered *hosia*.[52] Such land was either *idia* (for personal use, i.e., by particular households or groups of households) or *demosia* (for common use by the entire *demos*), with the great bulk of it falling in the former category, even if "the people" as a whole retained an ultimate proprietary control over all *hosia* holdings, as noted earlier.[53]

This same "*demos* of the Athenians" of course constituted the third component of the *polis* ecology. The primary contributions of "the people" to this reproductive system were not just to reproduce, preserve, and manage their own lives, but to defend and administer the ancestral land and the divine cults which had made all those lives possible in the first place. Yet more, much more, remains to be said about this *demos*, about how precisely the Athenians objectified themselves as a communal entity and agency. For as we shall now see, this self-understanding is substantially missing from our standard modern accounts.

The Personality of the Athenians

It was a commonplace among the Athenians and the Greeks in general that a *polis* was made of "men, not walls." It was a "communion" or "partnership" (*koinonia*) of persons, not an abstract political space or territorial formation. Thus, in the Greek mind's eye, "Athens" did not fight "Sparta." "The Athenians" always fought "the Spartans."[54] This much all modern observers readily see and acknowledge.

But what they tend not to see, because modern metaphysical commitments do not allow them to see it, is that this communion known as "the Athenians" was a unitary, free-standing agency in its own right. Unlike, say, liberal civil society, it was not merely an aggregate of pregiven individuals. It was a pregiven corporate subject, one that was not reducible to the particular living, breathing individual subjects who happened to comprise and embody it at any given time.[55] In the thought and practice of the Athenians, their *koinonia* was an ontologically autonomous thing-in-itself, a poliadic person or self that existed prior to and apart from themselves as discrete persons. As the human personality or essence of the *polis*, it was a kind of ageless primordial superorganism, one that had been continually present in Attica since the time of those first earth-born kings. As such, it had a certain subjectivity, a life, will, and an interest all its own. The Athenians called this living persona of their *polis* simply *ho Demos*, "the People."[56]

A well-known passage in Aristotle's *Politics* gives us a clearer sense of how this human personality of the *polis* was objectified and understood at the time:

> The *polis* is by nature clearly prior to the household and the individual, since the whole is of necessity prior to the part; for example, if the whole body is destroyed, there will be no foot or hand, except in the equivocal sense that we might speak of a stone hand; for when destroyed the hand will be no better than that ... The proof that the *polis* is a creation of nature and prior to the individual is that the individual, when isolated, is not self-sufficing; and therefore he is like a part in relation to the whole. But he who is unable to live in society, or has no need because he is sufficient for himself, must be either a beast or a god; he is no part of the *polis*.[57]

Which is to say, *polis*-dwelling Greeks did not think of societal membership in modern terms, in the universal, abstract, legalistic terms of "citizenship," as a bundle of institutionally protected rights, privileges, and obligations that bound an ever-changing cast of individuals to a group that was somehow external to themselves. They thought of it in altogether more immediate, more literal terms, as "having a share of" (*metekhein, meteimi*) the life of a particular, pregiven social body in a particular, pregiven land. As Martin Ostwald explains:

> To understand what is involved, we have to think of a "share" not in terms of the stock market, in which shares can be disposed of at will, but in the terms in which each limb has a share in the human body: my leg "shares" or "participates" in my body in the sense that whatever affects it affects my body, and whatever affects my body affects it.[58]

Hence, simply being born within Athenian territory did not make one an Athenian. Being born into the already existing "Demos of the Athenians" through Athenian parents, a mother and father who had themselves been nourished by the particular land and tutelary gods of Attica, was usually what made one an Athenian.[59]

Hence too, the constituent elements of this Athenian social body were *oikoi*, family or household units, not individuals. Like the cells of a human body, Athenian households existed in a symbiotic relationship with the compound organism which they all collectively comprised. Just as the existence of the social body as a whole depended upon the continued vitality of the household cells and their capacity to produce life-nurturing resources, so too no single household cell could subsist and produce such resources by itself if it were somehow abstracted from the social body.[60] To manage an *oikos* was thus to contribute to the management of the *polis*. The two were inseparable. Each presupposed the other, just as both presupposed the availability of a god-given land to sustain the life of the whole. So, in principle at least, each Athenian *oikos* was assumed to be an autonomous productive unit, responsible for extracting whatever means of life lay in its designated portion of the land of Attica, even as this autonomous productivity was itself only imaginable and sustainable through membership of the *polis*.[61] It was not the case, as one so often reads, that the "lines" between public and private "spheres" in Athens were "blurred." It was rather the case that there were no such discrete "spheres" there to speak of at all.

Evidently, in this particular metaphysical environment, there could be no room for any such creature as the free-standing, self-actualizing individual. So how exactly did the Athenians objectify personhood and selfhood?

Again, the prevailing ideas seem much closer to those of other non-modern worlds than they do to those of liberal modernity. Indeed, it may be helpful to think of the Athenian person as a kind of "dividual" self. Each Athenian, male and female, was simultaneously both a *polites/politis*, a constituent or expression of the social body of Demos, and an *idiotes*, a constituent or expression of a particular *oikos*.[62] In our modern terms, it is as if they were all two different people or beings at the same time, each one with its own, distinct form of personality or subjectivity. And what is important for our purposes is that both forms of subjectivity were essentially relational.[63]

That is to say, Athenians never stood just for themselves as disaggregated individuals, as singular instantiations of a universal personhood. They always confronted the world as constituents of a particular group or body larger than themselves, bearing the personalities of both the *polis* and the household that had made their lives possible in the place. They were not defined as persons by a bundle of supposedly natural, pregiven properties that were supposedly shared by all human beings everywhere, like the self-interest, the freedom, and the reason celebrated by modern liberalism. Their particular *polis* and their particular *oikos* was the essence of who they were. This meant that Athenians were not just culturally or politically distinct from others; they were essentially, ontologically distinct from all non-Athenians. *Oikos* and *polis* were the very materials they were made of, the very fabrics of their being. For them, there was no other kind of being, no other way to be in the world.

An Ecology of Gods, Land, and People

To help us further contextualize these distinctly non-modern forms of personhood and subjectivity and complete our preliminary survey of the metaphysics of Athenian community, we might briefly consider the contents of one of the more significant ceremonies in the *polis* calendar. I refer to the "entry rites" (*eisiteteria*) for "ephebes" (*epheboi*), eighteen-year-old male Athenian "cadets." This was the ceremony which marked the formal beginning of their two-year induction into full membership of Demos, a process which would require them to prove their claims to manhood precisely by guarding, patrolling, and, if necessary, defending the ancestral land.[64]

The setting for this critical rite-of-passage ceremony could hardly have been more auspicious or appropriate. It was staged on the eastern slope of the acropolis, an area of extraordinary historical and divine resonance. The site not only offered the participants a panoramic view of the land they were charged with protecting; it also placed them in the neighborhood of the "Cecropian agora," the location of the city's most ancient and hallowed communal buildings, like the Prytaneion, which housed Hestia ("Hearth"), the immortal life-fire of the *polis*.[65]

No less appropriate and auspicious was the immediate location of the ceremony. This was the sanctuary of the goddess Aglauros, a site dominated by a huge cave, which formed the most visible landmark on the eastern flank of the citadel.[66] A one-time mortal, Aglauros had been heavily involved in the very birth of the *polis*, as one of three daughters of Cecrops, the earth-born first king of Athens, as an actor in the nativity story of the similarly earth-born Erechtheus/ Erichthonius, Cecrops' successor, and as a heroic figure who had willingly sacrificed her own life to preserve those of her land and people during an early war between Athens and Eleusis.[67] Subsequently, Aglauros was worshipped at the site of her self-immolation on the east side of the citadel. Along with her sister Pandrosos and the Youth-Nurturing Land of Attica (*Ge Kourotrophos*) herself, she was venerated as a guarantor of Athenian vitality, both human and vegetal. And just as Pandrosos was assigned a special responsibility for the rearing of Athenian girls, so Aglauros came to be considered the patron divinity of male ephebes, whose annual initiations into Demos represented the regular renewal of the life of the *polis*.[68]

As for the particulars of the "entry rites" ceremony itself, we do not know which other Athenians were present, aside from the priestess of Aglauros and Ge Kourotrophos, who conducted the rites. But we do know that the occasion was attended by a distinguished array of gods, including many of those deities who performed the most essential, life-sustaining functions for the *polis*. Some of these were already at large in the area, like Athena on the acropolis, Hestia in the Prytaneion, and of course Aglauros and Ge Kourotrophos themselves. The rest were induced to attend through the *eisiteteria* offerings. And with this exalted divine company and the land of Attica thus assembled as witnesses, the central event of the ceremony could take place. The ephebes were duly invited to take up their "consecrated weapons" (*hiera hopla*) and swear the solemnest of oaths, vowing to act like true Athenians and defend all the contents of their *polis* to the death:[69]

> I shall not bring dishonor on these consecrated weapons (*ta hiera hopla*) nor will I abandon my comrade wherever I shall be stationed. I will defend both *ta hiera* and *ta hosia* and will not pass on my country (*patrida*) diminished when I die, but greater and better, so far as I am able by myself and with the help of all. I will respect the orders of all those who govern reasonably (*ton aei krainonton emphronos*), along with the existing ordinances (*ton thesmon hidrumenon*) and all others which may be instituted reasonably in the future. And if anyone seeks to destroy the ordinances I will oppose him so far as I am able by myself and with the help of all. I will honor the cults of my ancestors (*hiera ta patria*). Witnesses to this shall be the gods Aglauros, Hestia, War (female: *Enyo*), War (male: *Enyalius*), Ares and Athena, Zeus, Florescence (*Thallo*), Growth (*Auxo*), Fruitfulness

(*Hegemone*), and Heracles, along with the Boundaries of my native land,
Wheat, Barley, Vines, Olives, Figs . . . .[70]

And thus, from the sanctuary on the east side of the acropolis, this site of cosmic
convergence, between Athenians past, present, and future, between vital forces
human, terrestrial, and divine, the ephebes departed with their consecrated
weapons to assume their divinely appointed duty of preserving the life of the
polis, the social body that had made their own lives possible in the first place.

As moderns, so many worlds removed from this ancient ceremony, we can
hardly presume to grasp its affective force, to know what the young men of
Athens might have seen, thought, and felt as they prepared to cross the threshold
to manhood. But sources like the ephebic oath quite vividly convey what the
Athenians as a whole thought their real world was made of, offering us a glimpse
of their distinctly non-modern way of being human, of reality otherwise.

For theirs was a world governed ultimately by a small multitude of ageless, primordial beings, all with their own distinctive powers, histories, and subjectivities,
beings whose presence was largely unobservable, whose wills were all but unknowable, but whose actions impressed themselves visibly and continually upon
the fabric of material existence. It was a world whose contents were all thus, by
definition, either consecrated to gods (*hiera*) or acceptable to them (*hosia*). It was
a world in which the landscape itself and all its fruits were deemed to possess a
kind of personality and agency all their own, to work actively for the welfare of
their human protectors, to feel a certain kinship with those whose being they daily
sustained. And as such, it was a world where land, people, and gods were forever
entwined in a single, symbiotic communion, essentially bound to one another by
the very processes of life itself. This, broadly speaking, was the model or template
of social being that was actually lived and experienced by the Athenians, a model
that defies the most fundamental premises of our mainstream social science.

If so, we can safely abandon once and for all the idea of a "democratic Athens,"
not least because the whole metaphysical environment of the *polis* was manifestly
uncongenial to anything resembling a modern liberal democracy. The materialist,
anthropocentrist, secularist, and individualist foundations of western modernity,
the conditions of liberalism's very possibility as thought and practice, were simply
not there in antiquity.

So what exactly did *demokratia* mean in this particular non-modern lifeworld? In the book's remaining chapters, we shall continue this exercise in non-
dualist historicism by examining the Athenian *politeia* on its own ontological
terms, in its own metaphysical conjuncture, as an entire "way of life."[71] To be clear,
the immediate aim of this recursive analysis is not to produce an exhaustive alternative account of every single practice that somehow helped to sustain the life

of the *polis*. The goal is rather to recover the world-making common sense, the prevailing social knowledge or "science," which was realized in these practices, showing along the way how the fundamental logics, principles, and rationality of Athenian *demokratia* differed quite profoundly from those which animate our own liberal capitalist mode of existence today. And we can begin by considering in more detail probably the most obvious of these differences, namely the essential roles played by superordinate non-humans in the reproduction of Athenian social being.

Governed by Gods

AT LEAST ACCORDING to the highly influential account developed by Michel Foucault and elaborated by many of his followers, the production of order in modern liberal polities is the work of a historically novel mode of "government," with its own distinctive "governmentality." Here, the term "government" is used in an unusually broad sense to mean a generally non-coercive direction of behavior, a "conduct of conduct." It is a mode of rule that takes place whenever and wherever a scientifically informed "power-knowledge" is mobilized to align the behavior of subjects with larger societal objectives, conforming it to rationally determined norms of morality, vitality, productivity, and so forth. Instrumentalizing the findings of demography, economics, biology, psychiatry, psychology, and other modern sciences, this liberal government is thus exercised by all manner of specially trained, expert authorities at a vast network of different sites, from state ministries, schools, and universities to prisons, hospitals, clinics, and workplaces. In its most characteristic forms, it is a mode of rule that aims to transact a peculiarly modern "biopolitics," to take charge of and manage the very wellsprings of life itself.[1]

This brings us immediately to what must be the most fundamental difference between ancient Athenian and modern liberal approaches to life management. For in ancient Attica, humans did not presume themselves capable of controlling the wellsprings of their own vitality. The Athenian *politeia* took it for granted that gods were the ultimate managers of all lives and livelihoods in the *polis*. As noted above, at least two hundred different divine agencies were acknowledged in Athens during the classical era, and between them they governed all of life's essential conditions, processes, and outcomes.[2]

The Possibility of Life Itself

To an Athenian, it was self-evident that gods shaped and determined the conditions of all material existence. As supreme ruler of the skies, Zeus controlled all climates and weather patterns, dispensing thunder, lightning, and of course precious, all-important rain.[3] Gaia or Ge (Olympia), the earth goddess, held sway over all land and soil.[4] An array of other superhumans, including Poseidon, water nymphs, river gods, the Sun, and the Seasons, apportioned other basic requirements for life and flourishing, such as light, warmth, and moisture.[5] Then there were all those gods who had a kind of tutelary concern for particular crops and their life cycles. Demeter was responsible for grain yields, Dionysus for the growth of grapevines and the production of wine. Athena retained a personal interest in the many offspring of the olive tree she had once bestowed upon the Athenians. Apollo had the power to bring vitality or blight to corn and other staples, like nuts, berries, and beans.[6] Hermes traditionally oversaw the well-being of animal herds.[7] And still other deities, supernal and infernal, from Plouton, Ploutos, and Good Fortune (*Agathe Tukhe*) to Zeus Pankrates, Zeus Meilichios, and Zeus Philios, harbored a more general concern for the production and maintenance of agricultural abundance in Attica as a whole.[8]

By a similar division of labor, the gods apportioned among themselves responsibility for the different stages in the life cycle of the human organism. If a single divinity served as a kind of overseer of the reproduction of the *polis* community, it would be Mother Attica herself, as Ge Kourotrophos, to whom preliminary sacrifices were offered at many festivals across Attica, especially those generally associated with human life processes.[9] Various other deities, including Aphrodite, the Genetyllides ("Birth Spirits"), Demeter, Kalligeneia ("Fair Birth"), and Eileithyia, had more specific concerns for the processes of conception, pregnancy, labor, birth, and infancy.[10] Gods like Artemis Brauronia, Pandrosos, and Aglauros managed the all-important transitions from childhood to adolescence and beyond, while the Graces (*Kharites*) bestowed qualities like "Merriment" (*Euphrosune*) and "Splendor" (*Aglaia*) along the way.[11] And when the time came, still other gods, like Hera (Teleia), Aphrodite (Ourania), Artemis, and the Moirai ("Fates"), were responsible for the formation of successful marriages, while prospective spouses would seek help from the likes of the Tritopatores and the Semnai Theai to secure future offspring and wedded bliss.[12]

Gods also monitored and secured the fortunes of individual households. Zeus Ktesios, sometimes present in the form of a jar of *ambrosia*, watched over the wealth and possessions of the *oikos*, while Zeus Herkeios seems to have been responsible for preserving its integrity as a family unit.[13] No less essential to the well-being of the household was the continuing presence of Hestia, the hearth goddess, whose fire embodied the very perpetuity of its existence across generations, much as her flaming hearth in the Prytaneion in Athens manifested the eternal vitality of the *polis* as a whole. As the visible life spirit and physical center of the household, Hestia was thus the goddess to whom all new members of the *oikos* were first introduced, whether brides, children, or slaves.[14] Otherwise, protection of the *oikos* from external ills was furnished by a cohort of divine "averters of ills" (*alexikakoi*), like Apollo Aiguieus, Hermes, and Hekate, who performed apotropaic functions in entrance ways, usually in the form of aniconic or semi-iconic objects, like the well-known "Herm" statues.[15] As for protection and healing from more immediate bodily ills and sicknesses of all kinds, these services could be found in the world beyond the household, in the precincts of yet another diverse array of superhumans, including Asclepius, the hero Amphiaraus, Athene Hygieia, Aphrodite, Artemis Kalliste, and Apollo Paion.[16]

All this is to say nothing at all of divine involvement with manifold everyday human pursuits. Gods like Athena (Ergane) and Hephaestus, along with the hero Prometheus, nurtured and supported the domestic crafts of women, metalworking, and other essential productive skills and activities.[17] Seafarers might seek the help or favor of Poseidon Soter ("Savior"), Poseidon Pelagios ("of the Sea"), Aphrodite Euploia ("Pleasant Sailing"), or the Dioscuri, Castor and Polydeuces.[18] Traders might invoke support from a concerned divinity like Hermes Agoraios ("of the Market Place").[19] To secure successful outcomes for communal decision-making procedures, one could offer prayers and gifts to agencies such as Zeus Boulaios ("Counsellor"), Athena Boulaia, Artemis Boulaia, and Apollo Prostaterios ("Champion Protector").[20] And of course one could enlist the services of a small army of divinities, like Ares, Athena Areia ("the Warlike"), Athena Nike, Enyo, Enyalios, Artemis Agrotera ("the Huntress"), Apollo Lykeios, and Zeus Tropaios ("of the Battle Turning Point") to help with all manner of warfare-related activities, from the training of cadets to the celebration of victories.[21]

And this is still to say nothing of all those major gods who variously helped to nurture, protect, and perpetuate communal well-being and integrity across Attica as a whole. Zeus Phatrios, Athena Phatria, and Apollo Patroos safeguarded the fortunes of constituent descent groups within the *polis* and received cult honors at the Apaturia and Thargelia festivals.[22] Zeus Eleutherios, housed unusually in a stoa in the agora area, ensured that the Athenians were free from domination by others.[23] Nearby, the Mother of the Gods served as custodian of the various documentary records produced by Demos.[24] At the same time, the unity and integrity of this same Demos was preserved by Aphrodite Pandemos, who had apparently been venerated in Athens since the time of the original *synoikismos* of Attica, when Theseus first induced all the region's inhabitants to "live together" (*synoikein*) as one.[25] Then, looming over all these deities were the acropolis gods Zeus Polieus and Poseidon Erechtheus, whose continued support made the very existence of the *polis* possible in the first place.[26] And perhaps even a little above these two transcendent beings was of course Athena Polias herself, the ultimate patron of all Athenians, their divine "mother, queen, and guardian," who, in the famous image of Solon, "holds her hands above" to protect them from all harm.[27]

Overall, there thus seems to have been a general sense in Athens that the gods of Attica were, in the parlance of the time, working actively and continually for the "health and safety" (*hugieia kai soteria*) of the *polis* community.[28] One of the more vivid expressions of this communal article of faith comes in the prayer that was offered at the start of each assembly meeting and session of the council of 500. According to a compelling modern reconstruction, this prayer read roughly as follows:

> Let us pray to (various deities) that this meeting in the *ekklesia/boule* may go as well as possible, to the benefit of the *polis* of Athens and of ourselves personally, and that whoever acts and speaks in the best interests of Athens may prevail. Let us pray to the Olympian Gods, the Pythian, the Delian, and all the other gods, that if any one devises evil against the Demos of the Athenians, or plans to set up a tyranny, or to bring the tyrant back, or deceives the *boule* or the Demos of the Athenians, or betrays the *polis*, or takes bribes to speak against the interests of the Athenians, or debases the coinage, he and his *oikos* may come to a miserable end. But on the rest of us may the gods pour many blessings.[29]

Even from this highly selective and perfunctory survey, it is clear enough not only that Athens was indeed "full of gods," but that this diverse multitude of superhumans were the ultimate "governors" of the *polis*. They continually and actively performed all of the most essential life management functions in Attica, including the kinds of functions that we might today call "biopolitical." Unlike ourselves, the Athenians could never have imagined a world where humans might in any meaningful sense take charge of the sources, means, and ends of life itself, the very conditions of its possibility. Which is to say, they could not have conceived of life in a modern secular metaphysical environment.

Ritual Reproduction of the Conditions of Existence

It follows that the principal human contributions to vitality in the ecology of the Athenian *polis*, the most essential, life-sustaining practices in the *politeia*, had little or nothing to do with what we conventionally think of as "politics" or "government" at all. On the contrary, if there was a human art or science of *polis* management in Attica, it began with the performance of rituals. For once we come to recognize the reality of divine governance in Athens, we can then also see that by far the most vital business transacted by the Athenians was not what they did in the assembly, council, courts, and other such "governmental" institutions. It was, rather, the business they transacted continually at a multitude of shrines and sanctuaries, the business of transforming *hosia* into consecrated *hiera*, the endless essential business of engaging with gods in ways that might encourage them to secure the "health and safety" of the *polis* as a whole. Ritual activities were not the routinized expression of some simple-minded, irrational faith, belief system, or "religion." When understood in their own metaphysical conjuncture, they were highly rational devices of a most urgent, far-reaching ecological significance. They were nothing less than the mechanisms used to reproduce the very conditions of existence.

The extreme ecological gravity of the stakes here in turn helps to explain inter alia why ritual activity was so continuous and ubiquitous in Athens; why little expense was spared to produce suitably lavish sanctuaries, temples, sacrifices, dedications, festival processions, and other ritual spectacles, especially for the major gods of the *polis*; why the annual passage of time itself in Athens was marked by a monthly calendar that was structured around the veneration of divinities; why at least three items of business concerning *ta hiera* of the *polis* had to be discussed at half of the forty regularly scheduled assembly meetings in any given year;[30] why the performance of all prayers and rituals in Athens was always marked by an extraordinarily scrupulous fidelity to the traditionally prescribed words and actions; why those convicted of harming the *polis* in some way, whether male or female, were commonly banned from entering all Athenian sanctuaries; why the misperformance of rites, the introduction of unauthorized new gods, the theft of temple treasures, and other displays of *asebeia* ("irreverence") towards *ta hiera* and the gods of the *polis* conventionally warranted a sentence of death or exile from all Attica *sine die*.[31]

Viewing ritual activities as mechanisms of ecological reproduction rather than mere expressions of "religion" alters our perspective on other matters too. For example, it allows us to see that the various bodies of knowledge which guided Athenian engagements with divinity were the functional equivalents of the various "sciences" which inform modern biopolitics. The cosmologies, the aetiologies, the memories of divine interventions and benefactions past, and all the other narratives that were reenacted in or presupposed by ritual performances were not simply time-honored "traditional tales" or sources of mere entertainment, ancient "myths" that were dutifully remembered and retold merely because "custom" or "the ancestors" had so ordained. These stories were remembered and retold because they contained timeless truths about the origins, causes, and nature of things, about the mysterious ways of gods to humans, and about the human devices that had successfully secured divine favor and the welfare of the *polis* in times past.

Likewise, knowledge about the appropriate ways to engage with particular deities, knowledge about how, when, where, and by whom offerings should be made to specific gods, was not transmitted from one generation to the next from sheer force of communal habit. This knowledge was scrupulously preserved and reproduced, whether by oral or written means, because its efficacy had been conclusively proven in practice. Like the findings of modern biology, climatology, epidemiology, psychology, dietetics, and all the other sciences upon which our capitalist way of life is staked, Athenian ritual knowledge retained an unimpeachable, "scientific" truth status because its contents had repeatedly been tested and verified by experience, above all by the self-evident survival and flourishing of the *polis* itself since time immemorial.[32]

Moreover, recategorizing ritual practices as exercises in ecological management significantly alters the way we think about the human "managers" of life in Athens. After all, the performance of rites that sustained the very fabric of the *polis* and all its human and non-human contents was not effectively monopolized by a cohort of professionally credentialed experts or specialists, as the equivalent functions would be in, say, a Roman Catholic order. Nor was it even restricted to those who were entitled to attend and vote in the assembly meetings. Indeed, to a point, one could say, the responsibility for the ritual reproduction of life's most essential conditions was dispersed among all humans in Attica, both males and females, rich and poor, young and old, native and non-native, free and unfree alike. Thetes, children, metics, and slaves all routinely took part in ritual activities.[33]

But it is the prominence of females in these ecologically vital activities that should perhaps draw particular attention. Of course, this prominence has long been observed and discussed by historians of ancient Greece.[34] But it is usually seen as something like an exception that proves a more general rule, as a kind of anomalous concession to oppressed females by patriarchal males, who begrudgingly allowed their wives and daughters out of the house to play a token "public" role in the materially inessential, make-believe realm of "religion," while they themselves got on with the business of managing "real life" elsewhere, in the assembly, council, courts, agora, and so forth.[35] This, I think we can safely say by now, is an unhelpfully ahistorical, modernist perspective. Females were routinely, sometimes exclusively, entrusted with administering and performing rituals that were absolutely essential to the "health and safety" of themselves, their *oikoi*, and their *polis*.[36]

The Life-Sustaining Ritual Responsibilities of Females

This remarkable ritual prominence speaks to a general perception among the Greeks that gods felt a special affinity with human females. This perception is succinctly expressed in a passage that is widely thought to come from Euripides' lost tragedy *Melanippe Captive*, where a female speaker defends the value of women to the *polis*:

> [Women] manage households (*oikous nemousi*) and take care of things brought home by sea. A household without women is a household without cleanliness and prosperity. Moreover, regarding engagements with the gods, matters which I judge to be the most important of all, we women have the greatest part to play (*meros megiston*). For example, in the house of Phoebus Apollo, women interpret the god's intentions, and at Dodona's august shrine, by the consecrated oak, womankind conveys the plans of Zeus to all Greeks who wish to know it. Women also perform all the rites for the Fates and Nameless Goddesses (i.e., Semnai Theai), rites which cannot legitimately be performed by men but which flourish in women's hands. In this way, a feminine order (*dike theleia*) prevails in human engagements with the gods.[37]

Female ritual responsibilities in Athens thus began at a very young age. Four such duties are mentioned in a well-known speech in Aristophanes' *Lysistrata* (411 BC), where the Old Women of the Chorus defend their claims to offer sound advice to Demos in wartime by invoking their various vital services to the *polis*:

> To all you city-dwellers (*pantes astoi*), I say, we are now going to offer useful advice for the *polis*. And rightly so, since the *polis* splendidly nurtured me in some comfort. Upon reaching the age of seven, I served right away as an *arrhephoros* ("bearer of unnameable objects"). As a ten-year-old, I was an *aletris* ("grain grinder") for our divine leader [Athena]. Then I donned the saffron robe as an *arktos* ("bear") at the Brauronia. And then as one of the beautiful girls, I served as a *kanephoros* ("basket bearer"), wearing a necklace of dried figs. So I have earned the privilege of offering useful advice to the *polis*. Nor should my womanly nature cause you to begrudge me this privilege if I say something advantageous to the present situation. After all, I am well able to make my own contribution to the common cause. I contribute men.[38]

Of the various cult practices mentioned here, the most significant was probably the *arrhephoria*, which was performed annually by four girls between the ages of seven and eleven, all of them from elite families and all specially chosen by the *archon basileus*, one of the nine archons. Apparently, these *arrhephoroi* lived for a time with Athena herself on the acropolis, helped with the weaving of the goddess's annual new robe (*peplos*), and were responsible for "carrying the unnameable objects" (*arrhephorein*) on an appointed night from Athena's precinct down through a secret, underground passage to the sanctuary of Aphrodite in the Gardens, whence they subsequently returned carrying some similarly mysterious, similarly hallowed objects.[39]

Otherwise, it seems that the *aletrides* were responsible for grinding grain into small consecrated cake offerings for Athena, but little more is known about their activities.[40] The peculiar practice of the *arkteia*, where young girls "acted as bears" (*arkteuein*) at the sanctuary of Artemis Brauronia on the east coast of Attica, was remembered by tradition as an expiatory rite, an annual attempt to assuage the wrath of one of the region's most significant deities following the killing of one of her bears. The more lavish quadrennial version of the Brauronia festival, of which the *arkteia* was a part, was considered sufficiently vital to the life of the *polis* that it was included in the small group of festivals that were expressly administered by a board of ten lottery-appointed *hieropoioi* (literally, "sacrifice makers").[41] Meanwhile, specially chosen, unmarried females served as *kanephoroi* at a variety of festivals, including the Panathenaia itself. It appears that they commonly led the ceremonial procession, since their baskets contained the sacrificial knife itself and other consecrated items.[42]

And of course, this ritual prominence of females continued long after girlhood. It was most obviously visible in the activities of those chosen to serve as the direct human intermediaries with divinity, namely the priestesses of the *polis*.[43]

While priests and priestesses may ultimately have been "servants" of the community, their particularly essential form of service nonetheless gave them a special standing among the people, as "the Stranger" in Plato's *Statesman* remarks to Socrates:

And then again there are the priests who, custom tells us, have an understanding that allows them both to convey gifts to gods through their sacrifices and to request from them with prayers the benefits that we possess.... [T]he bearing of priests and prophets is very much full of dignity and they acquire an august reputation (*doxa semne*) because of the magnitude of their undertakings.[44]

In classical Athens, males and females were equally eligible for this particular form of prestige, since, as a general rule, male deities had to be served by male priests and female gods by priestesses.[45] And this meant that women were directly responsible for a good number of the most ecologically consequential ritual actions that a human in Athens could ever perform, given that priestesses served many of the most significant cults in the *polis*, including those of Demeter and Kore at Eleusis, Artemis Brauronia, Athena Nike, the Semnai Theai, and of course Athena Polias, the divine protectress of Athens herself. So it is no surprise that some of the more distinguished holders of these illustrious positions acquired a certain measure of celebrity during their lifetimes and sometimes thereafter.[46]

More generally, female members of the *polis*, young and old alike, were the primary or exclusive participants in various ritual events upon which the life of the community as a whole was in some sense staked. In addition to the *arrhephoria* and *arkteia* mentioned above, females were solely responsible for performing the requisite rites at a number of Athenian festivals, including the Adonia, the Skira, and the Stenia, each of which in some way concerned the general fecundity of the homeland.[47] Even more significant than any of these was the Thesmophoria, which is generally regarded as the principal celebration of "fertility," both human and non-human, in the Athenian calendar. Presided over by the priestess of Demeter and Kore, managed by a pair of female "rulers" (*arkhousai*), and lasting for three days, the festival encompassed an intriguing mixture of elements, including the temporary residence of the celebrants in Demeter sanctuaries, fasting, *aiskhrologia* ("speaking shameful things"), the depositing of piglet remains in special pits, and a banquet. As Eva Stehle has observed, the net outcome was an event where "women intervened as agents in the cosmic order and confirmed their collective power while experiencing an intimate encounter with a goddess."[48]

Last but not least, one can hardly overlook the prominence of females, native-born and metic, in the single most important ritual activity in Athens, namely the Panathenaic festival itself. Most conspicuously, in the grand procession (*pompe*) up to the acropolis which preceded the principal sacrifice, the front ranks of the processants were filled almost exclusively by girls and women. According to Lisa Maurizio's plausible reconstruction, which draws on evidence from the famous Parthenon Frieze and other visual and literary sources, the first eight groups were as follows: Athenian girl basket-bearers (*kanephoroi*); metic girl parasol-bearers (*skiadephoroi*); metic girl stool-bearers (*diphrophoroi*); Athenian female weavers of the *peplos* (*ergastinai*); Athenian girls who carried the "unnameable objects" (*arrhephoroi*); sacrificial animals; metic male tray-bearers (*skaphephoroi*); and metic girl water-bearers (*hydriaphoroi*).[49]

In other words, as bearers of most of hallowed ritual paraphernalia, as leaders of the procession, and thus as the humans who were in nearest proximity to the goddess herself, females were "associated more closely with the divine" and assigned the most "prestigious" and "honorable" roles in the event.[50] At the most significant festival in Athenian experience, where the special relations which bound a people to its chief divine patron were annually renewed, girls and women bore the primary ritual responsibilities. Again, it seems, females were readily entrusted with essential, life-sustaining functions, functions upon which the very existence of the *polis* self-evidently depended.

Of course, this routine prominence of females in ritual activities creates serious difficulties for modern conventional wisdom about the production of order in classical Athens. As Maurizio acknowledges, the very contents of a festival like the Panathenaia resolutely defies our persistent attempts to read it as a "political" event, one that somehow exemplified or instantiated the male-dominated "democracy" which most believe defined the essential nature of the *polis*.[51] But if, as I am recommending, we instead see *demokratia* on its own ontological terms, as an altogether more comprehensive, more inclusive "way of life," one that obliged all members of the *polis* to help reproduce a cosmic ecology of gods, land, and people, then such difficulties soon start to disappear.

For once we embrace this alternative, recursive approach, recognizing above all how the management of life in Attica depended ultimately and unavoidably upon ritual interactions with divinities, then we can start to appreciate more fully how the world of the ancient Athenians was quite fundamentally and categorically different from our own. In the end, there simply can be no real resemblances between worlds formed under such radically different metaphysical conditions. There can be no meaningful similitude or commensurability between an environment where superhumans alone could secure all the basic fabrics of existence and one where humans are presumed to be capable of taking charge of life itself.

Needless to say, the essential differences between ancient Athenian and modern capitalist ways of life hardly end there. Over the next four chapters, we will deepen our sense of the ontological and metaphysical distance between the two as we consider how humans managed their own affairs in Attica.

12

The Cells of the Social Body

TO BE HUMAN in our current conjuncture is to have the contents of one's life constantly monitored, measured, and otherwise managed by innumerable largely unseen professional authorities, by specialist agencies who often operate in ways that one can barely understand, never mind control. Some function within the ambit of the public sector or state, administering on our behalf matters of law, police, defense, taxation, immigration, naturalization, transportation, education, environmental protection, social security, and so forth. Many others function partly or wholly outside this public realm, such as the managers of healthcare networks, financial institutions, insurance companies, telecommunications services, and human resource departments in private workplaces. But all such agencies, public and private alike, are deemed somehow essential to the continued vitality of our current lifeworld. All work incessantly to maintain and reproduce the conditions of modern existence, to maintain an ecology congenial to capitalist flourishing. Without all these superordinate agencies of "government," it seems, our cherished free societies, with their life-nourishing free markets, would be unsustainable.

The ancient Greeks were themselves all too familiar with government by largely unseen powers, powers who moved in often mysterious ways. But as we have just seen, these authorities were exclusively divine, not human. When the Athenians claimed that in Athens one was free "to live as one liked," they meant that they were not ruled by other human persons, by specialist, superordinate mortal "governors." They could not have imagined modernity's experience of continuous government by human experts, because they did not know themselves as a free-standing "society," as an entity that was governable from without by other similarly free-standing entities, whether public, private, or otherwise. They knew themselves rather as a consanguineous people, a social body of households that were all individually and collectively responsible for governing themselves.

In the following three chapters, we shall look in turn at the three different types of human agency that were responsible for managing life in Attica: the single, autonomous *oikos*; groups of *oikoi*, bound by ties of blood, cult, place, and friendship; and the unitary body of all Athenian *oikoi*, the "Demos of the Athenians," which made and enforced rules for the *polis* as a whole. We shall look first at the individual *oikos*, which was probably the most important of the three. For the Athenians simply took it for granted that most of life's more essential, more vital concerns would be administered on a day-to-day basis by parents in households, with little or no help from others. But before we can fully appreciate the importance of household management, we must first reconsider both the nature of *oikos* and *polis* as entities and the relations between the two.

A Polity Composed of Households

In the one detailed, extended "theoretical" analysis of *polis* order that we possess from classical Greece, the author could not be more clear on the matter. "The *polis* is composed entirely of households" (*pasa . . . sunkeitai polis ex oikion*), each one of which "is a partnership (*koinonia*) established by nature for the satisfaction of everyday needs." And building on these premises, the work we conventionally know as Aristotle's *Politics* then proceeds to devote the whole of the rest of Book 1 (chapters 3–13) to the business of *oikonomia*, or "household management," discussing topics like the relations of husbands to wives and fathers to children, the management and education of slaves, and the acquisition and handling of essential material resources.[1]

And to a point, historians have come to recognize the significance of *oikoi* to the life of the *polis*. More than a quarter of a century ago, Lin Foxhall began an exceptionally insightful and valuable discussion of households and gender relations in classical Athens with the following words:

> The idea that the household was the fundamental building block of ancient Greek society, explicit in the ancient sources, has now become widely accepted. It is no exaggeration to say that ancient Athenians would have found it almost inconceivable that individuals of any status existed who did not belong to *some* household; and the few who were in this position were almost certainly regarded as anomalous.[2]

But as Foxhall goes on to point out, specialists still for some reason continue to assume that "the foundations of power" in Athens and other *poleis* "lie solely in the public sphere, and that domestic power is 'less important.'" Challenging this assumption, she then proceeds to make a compelling case that "the household" was never "outmoded" or "superseded by political institutions" in classical Athens. On the contrary, in areas like "property holding" and "citizenship," political and domestic roles were "intertwined with each other." Every Athenian "still belonged to both *polis* and household, and the bonds to both were strong."[3] And in the time since Foxhall's paper was published, a good number of other studies have reached similar conclusions, broadening further our sense of the importance of *oikoi* as the "building blocks" of the *polis*.[4]

Yet as far as one can tell, this kind of work has so far had at best only a modest impact on orthodox accounts of the Athenian *politeia*. Indeed, one cannot help but notice that the last quarter of a century has also witnessed the extraordinary efflorescence of "Athenian democracy studies" as a more or less self-contained field. Over this time, there has been a veritable explosion of research on political institutions and ideologies, law, citizenship, civic identity, and the possible articulations between "democracy" and just about every other facet of Athenian experience, from rituals and athletic spectacles to art, architecture, theater, and sexuality. And in general discussions of the basic fabrics of Greek experience, one still routinely encounters statements about *polis* and *oikos* which quite explicitly subordinate the latter to the former, treating the two as if they were entirely separate, even rivalrous entities that somehow competed for the loyalties of their constituents. Here, for example, is how one well-regarded specialist work on life in Athens introduces the topic of "the *oikos*" to its readers:

> Between the citizen and the state lay the *oikos*, variously translated as "the family," "the estate," "the household," "the House." In Athens it found itself in a rather precarious position, squeezed between the stronger claims of the individual and the polis. For the special relationship between city and citizen which is such a defining feature of classical Greece in general and democratic Athens in particular could only be achieved by damping down other affiliations that might get in the way.[5]

I suggest that the reason for this continued marginalization of the *oikos* in mainstream modern accounts of the *politeia* has nothing to do with any perceived weaknesses in the arguments made by the likes of Foxhall. Rather, it has everything to do with the basic analytical template which we use to render Athenian social being intelligible in the first place.

For, as the passage just quoted well illustrates, this template obliges us to reengineer the Athenian lifeworld at the ontological level, translating a social body that was "composed of households" into a modern-style, dichotomized societal terrain, where a private social realm of "the *oikos*" now stands apart from a free-standing public political realm of "the *polis*." In this characteristically liberal vision of order, "real power" (i.e., a juridico-coercive sovereignty) will by definition be concentrated in the political field, whence it radiates out over the social field, constraining the liberty of individual subjects. So it is all but inevitable that "the *oikos*" will come to seem subordinate to "the *polis*" in our modern accounts. Thus, however much we might insist that the *oikos* was "important to the Greeks," this importance will always be denied or diminished by our own ontological presuppositions. Our standard template of social being simply cannot make sense of a polity that was "composed of households," a polity that defies our very notions of public and private.

Hence, to produce more historically meaningful accounts of the Athenian *politeia*, we need to abandon our modernist preconception that *polis* and *oikos* must have occupied separate bounded fields in a functionally compartmentalized societal terrain. We must instead see Athens as it saw itself, namely as a unitary body of household cells. We must effectively reeducate ourselves to see that Athenian *oikoi* constituted the very fabric of the Athenian *polis*. Thus, to be a member of the *polis* was to be always already a member of one of its constituent *oikoi* and *vice versa*. And if we are prepared to take seriously this essential symbiotic interdependency between the two, our perspective on Athenian order production, on the management of life in the *polis*, will be quite significantly altered.

Most immediately, we will see that much of the fundamental business of managing the life of the *polis* was actually transacted by individual households. For in a world that knew nothing of any discrete "public realm" or "domestic sphere," to manage an *oikos* was, by definition, to contribute to the management of the *polis* as a whole. Moreover, in a world which knew nothing of any abstract, freestanding "system" of "economy," capitalist or otherwise, the *oikos* was the primary means of life itself for its members. There were no real alternatives. Yes, one could supplement one's resources with stipends earned by, say, serving as a councillor or judge, rowing in the navy, or working on *polis* building projects. But there were no companies, no businesses, no organizations out there from which one might earn a stable, long-term income. Nor would one work routinely for the household of another, since it was slavish to depend on someone else for one's own livelihood. The reason why households were so "important" to the Greeks was not because of some traditional, "pre-political" attachment to "the family." Households were important because they remained the essential means of subsistence itself, for both rich and poor alike. So in a very real, very immediate sense, the business of "government" in Athens began with the management and perpetuation of each *oikos*.

And this alternative way of viewing Athenian "government" in turn prompts us again to reconsider fundamentally the contributions of females to the *politeia*, given that women were expected to be the primary "managers" of households at the time. We noted back in Chapter 2 how Athenian women were recognized at the time as *politides*, as "female members of the *polis*," even if conventional wisdom about the prevailing gender regime would question the possibility of any such membership. As we shall now see, viewing the management of life in Attica from the proposed recursive perspective will help us to make altogether better sense of the term *politis*, allowing us to restore to it some of its original force and content.

Female Managers of the Polis

Of course, there were imbalances between husband and wife. In a world where inheritance was patrilineal, where a wife would typically marry into a pre-existing *oikos* from without, it is simply unrealistic to expect that she could normally have become the titular "head" of that household, its principal "face" and agent in the outside world. Of course, in a lifeworld that was far more precarious than our own, there was usually a significant imbalance between the ages of husband and wife, given that the former had to be old enough to assume the responsibilities of household custodian (*kurios*), while the latter had to be young enough to maximize the chances of producing living, legitimate heirs. Of course, more generally, it seemed at the time that nature and custom had ordained a highly gendered division of societal labor in Athens, whereby males and females were assigned distinctly different roles in the reproduction of the social body. And of course the specific role of household *kurios* was performed by a male under normal circumstances. But this same division of labor did nonetheless make the wife responsible for administering most if not all of the indoor tasks that were essential to the everyday life of *oikos*.

Again, the ancient sources which explicitly address this issue are quite unambiguous. Like so many other non-modern peoples, the Greeks took it for granted that the gods had quite deliberately engineered male and female to serve different, if complementary roles in the work of perpetuating the lives of households and communities. In Greek thought, humans were not generic, presocial, interchangeable individuals, all naturally equal and naturally free to pursue their own self-realization as beings. They were expressly created from the start to perform particular gendered functions in the lives of the particular social bodies to which they would naturally and inevitably belong.

In perhaps the best known source on the subject, Ischomachus, the heir to a large family estate, recounts to Socrates the way he had explained to his new, much younger wife how the gods ordained a precisely gendered division of societal labor:

> For it seems to me, my wife, that the gods exercised extraordinary thoughtfulness when they ordained this coupling (*zeugos*) between what we call male and female, seeing to it that it be most beneficial both to the couple

themselves and to their community. First, in order that the races of living creatures be not extinguished, this coupling is brought together for the production of children. Second, from this coupling, a source of mutual support in old age is provided to creatures, to humans at any rate. Third, unlike that of other creatures, who dwell in the open air, the way of life for humans clearly needs to be conducted under rooves.[6]

And as Ischomachus goes on to elaborate, it is this peculiar human need to live in roofed dwellings that most immediately explains which particular qualities and properties the gods assigned to males of the species and which to females. While the former were created expressly to perform the more physically demanding outdoor tasks like ploughing, sowing, and harvesting, the latter were designed specifically for the complementary indoor tasks, like keeping the fruits of agrarian labor secure in the house, rearing children, baking bread, and weaving clothes from wool. God duly endowed males "in body and soul" with a "greater capacity to endure heat and cold, journeys and expeditions," while he equipped the bodies and souls of females with a naturally greater measure of affection for newborn children and a greater measure of fear, since "a fearful soul is no worse at guarding." At the same time, since qualities like memory, diligence, and continence were required by all humans, he assigned them to both males and females to an equal degree.[7] Man and woman were thus designed by gods and encouraged by laws to be complementary, mutually beneficial "partners" (koinonoi) in the task of household management, each one at once depending on the other's strengths and compensating for the other's deficiencies.[8]

While the male, as the household's primary provider, defender, and representative in the world beyond, could be said to be its natural "leader" (hegemon), Ischomachus then reassures his wife that the female also had a no less significant role to play. Much like "the leader (i.e., queen) of the bees" (he ton melitton hegemon)

> it will be necessary for you to remain indoors. And those servants whose work is outside, you must send outside, while the labor of those who have to work inside must be directly supervised by you (epistateon soi). You must receive whatever is brought into the household and distribute what must be expended. And as for what must be set aside, you must use foresight and discretion, lest you spend in one month the expenses that are meant to last a year. When wool is brought you, it is your task to make sure that clothes are made for whoever needs them. It is also your task to make sure that the dry grain is good quality and fit for consumption. . . . And whenever any of the servants become ill, it is your task to see to it that they are nursed.[9]

And finally, as Ischomachus is keen to stress, the rewards and consolations for taking on all of these essential responsibilities could be considerable:

> My wife, other personal concerns will come to be pleasurable for you, such as when you take someone who is unskilled at spinning and make them skilled, thereby making her twice as valuable to you. So too when you take someone who is unskilled at accounting and attending others and make her a knowledgable, trustworthy, and attentive servant who is then worth infinitely more. So too when you have the power to treat well those who are moderate and beneficial to your household and to punish those who prove to be wicked. But the most pleasurable thing of all is that if you prove to be a better person than me and make me your servant, you have no need to fear that you will have any less honor with advancing age. On the contrary, you can be sure that, as you grow older, the better a partner (*koinonos*) you are to me, and the better a guardian (*phulax*) you are to the children of the *oikos*, the more honored you will be in the *oikos*. For the things that are noble and good are multiplied (*epauxetai*) for human beings not through a natural process of ripening but through the active cultivation of life-sustaining virtues (*aretas eis ton bion*).

Lest one might think that there is something somehow anomalous about Ischomachus's vision of marital "partnership," where the wife is purposely equipped by the gods to serve as an essential "co-leader" in the management of her husband's patrimonial household, much the same basic vision can be found in the work of Aristotle, who is not usually noted for his "progressive" views on gender relations. Indeed, he is on record as claiming that "male" (*to arren*) is by nature the ruler of "female" (*to thelu*).[10] But apparently this belief did not rule out the possibility of females playing an active and productive role in the management of households. As we learn elsewhere, Aristotle also recognized that there was a naturally/divinely ordained complementarity between male and female which equipped them to function as "partners" in life:

> . . . [N]othing is more natural than the partnership (*koinonia*) between male and female. For we have elsewhere established the proposition that Nature is intent upon multiplying her various works, not least each of her living creatures. And such multiplication is impossible for a male to accomplish without a female and *vice versa*. So the partnership between them is established by necessity. . . . And so with this [reproductive] purpose in mind, the natures of male and female have been preordained by Divine Providence (*prooikonometai hupo tou theiou*) to function in partnership.

For they are distinguished from one another by the possession of a faculty (*dunamis*) that is not adapted in every case to the same tasks, but in some cases for opposite ones, though contributing to the same end.[11]

And much like Ischomachus earlier, Aristotle then goes on to enumerate the salient differences between the complementary natural "faculties" of males and females.

For Divine Providence made the man physically stronger and the woman physically weaker. Thus, the male, by virtue of his manly courage, is better able to defend the household and furnish it with resources from without, while the female, by virtue of her timidity, is better equipped to guard the household and secure its resources within. Likewise, in the area of domestic handicrafts, woman has been endowed with a sedentary patience, but is weak when it comes to outdoor exertions, while the male is less equipped to cope with quietude, but fitter for physically active pursuits. In the production of children, both share alike, but each makes a distinct contribution to the upbringing. While the mother nurtures, the father educates.[12]

In real life, of course, things could be more complicated. In one of the better known forensic speeches in the corpus of Athenian oratory, one Euphiletus attempts to justify his killing of an alleged philanderer called Eratosthenes on the legal grounds that the latter had seduced and thereby "corrupted" (*diaphtheiretai*) his young wife. But when describing the early days of his marriage, Euphiletus nevertheless shows us further how a typical "partnership" between an older man and younger wife might evolve in practice.

For when I decided to marry and had introduced a wife to my household, for some time I was disposed neither to vex her nor to leave it to her too much to do as she pleased. I watched over her when I was able to and kept my attentions within reasonable limits. So when a child was born to me, I already trusted her and handed over all my affairs (*panta ta emautou*) to her, believing that there was now a very great intimacy (*oikeioteta megiste*) between us. Indeed, in the early days, fellow Athenians, she was the best of all wives, by which I mean she was a skilled, thrifty, and excellent household manager (*oikonomos*) who organized everything (*panta dioikousa*) very precisely.[13]

And it is not hard to find scattered elsewhere across the sources references to the specific items of household business which might routinely involve some significant level of female "management" and "organization." In addition to regular concerns like food preparation, cleaning, weaving, child-rearing, ritual activities, and training and supervising slaves, these included, in no particular order: the disposition of household finances; tending to the sick and the deceased; participating in the family council; helping to resolve disputes between family members; visiting and supporting imprisoned relatives; and arranging adoptions and marriages. Moreover, females commonly served as sources of knowledge and information about the history, resources, problems, and general operation of their households. As Virginia Hunter has stressed, such evidence consistently shows that women were "at the center of family decisions, being consulted and in turn having influence over others."[14]

Of course, all this is still to say nothing of the many and various different *oikos*-sustaining occupations in the "outside world" which poorer Athenian wives and mothers might also take it upon themselves to pursue, with or without help from their husbands. As far as we can tell from the evidence, such "professions" included but were not limited to: wet nurse; dry nurse; midwife; obstetrician; groom; weaver; tailor; fuller; seamstress; washer; cobbler; gilder; inn-keeper; farm worker; maker of garlands, ribbons, and other accessories; and purveyor of bread, fruit, vegetables, beans, lentils, garlic, sesame, salt, honey, porridge, and other foodstuffs, raw and cooked.[15] Indeed, given the relative profusion of this evidence, along with current opinion about the highly uneven distribution of land among Athenian *oikoi*, it was perhaps only in relatively prosperous families that a wife's contribution to the life of her household would have been limited to the daily business of *oikonomia*.[16]

Modern Translations of Ancient Female Experience

Again, by analyzing the kinds of evidence discussed above more on their own ancient terms, a growing number of historians in recent years have seen fit to challenge the rather bleak accounts of female experience that one still commonly encounters in the literature on classical Athens, emphasizing the essential significance of the many different contributions made by women to the Athenian way of life.

For example, in her ground-breaking work on the "policing" of Athens, Virginia Hunter has stressed the active, productive roles played by women in the everyday management and regulation of *oikoi* and neighborhoods, with some females even serving effectively as "heads of households" (*kuriai*) in certain circumstances.[17] Likewise, in the work cited a little earlier, Lin Foxhall has encouraged us to see that women were routinely involved in the disposition of land and other household resources, since husband and wife "had to act in consensus," functioning in practice as "a single social entity" when attending to such consequential matters.[18] Elsewhere, the same author has also shown how females were able to work through others to secure courtroom verdicts that were beneficial to themselves and their *oikoi*.[19] Otherwise, Edward Cohen has called attention to women's "widespread involvement" in the "business activities" of households, especially in activities like retailing, manufacturing, and banking.[20] And at a more fundamental, more structural level, Cynthia Patterson has in a series of important studies highlighted the critical contributions made by women to the integrity and vitality of the *polis* as a whole through their roles as wives, mothers, and household managers.[21]

Taken together, such studies help to enrich our sense of the valence of the term *politis* by delineating the possible contents of a female "membership" of the *polis*. But again, it appears that conventional wisdom has yet to fully digest the implications of such work and accommodate them within our standard accounts of life in "democratic Athens" and elsewhere in the Greek world. All too often in these accounts, it seems, historical nuances, complexities, and contingencies are obscured and occluded by a kind of modern metanarrative, one that insists on reducing all non-modern gender relations to a perpetually and inevitably one-sided power struggle, as "patriarchal" males devise ever more effective ways to assert their dominion over their largely inert, mostly docile female victims.[22]

For example, when it broaches the topic of gender relations for the first time, one widely used history textbook sees fit to warn its student-reader up front that "misogyny is an important theme in Greek culture," with such statements inevitably coloring that reader's understanding of all the "history" that then follows.[23] In similar fashion, a rival textbook introduces its rather brief discussion of the relations between *oikos* and *polis* with the unqualified assertion that "Greek society was male dominated, in a word, 'patriarchal.'"[24] Likewise, in one of the most widely used handbooks on "women in ancient Greece," the author states at the outset that her express intention is to contribute to "the history of the subordination of women," apparently taking it for granted that "domination" by males was the central, defining fact of the Greek female condition.[25]

To take another example almost at random, this lingering tendency to define female experience in antiquity primarily by negatives, by oppressions, denials, exclusions, and so forth, is still quite evident in a prominent, multi-authored "companion" to "women in the ancient world" which was published as recently as 2012.[26] Thus, in the signature essay on women's relations to "the state" in the archaic and classical eras, one finds it confidently asserted that "Greek thinking on woman began by excluding her conceptually from man and proceeded to exclude her socially, legally, and culturally from male social/political institutions."[27]

More assertive still is the essay on "women and law," where we learn not just that women in Athens "were hardly citizens in the proper sense of the word," but even that women in antiquity could claim only a "marginal status" in "the category of the human." As the author goes on to add for good measure, "the 'protective' (i.e., 'weakness') argument" which ancient men used to justify their treatment of women was "bogus," being merely "designed to enshrine the rights of male heads of families and their control of property for themselves and the male descendants who gave them and their house immortality."[28]

Clearly, these mainstream accounts of gender norms in Greek antiquity allow little room for any meaningful female "membership" of the classical Athenian *polis*. And again, I suggest, the ultimate reason for the persistence of such questionable conventional wisdom is not to be found in the analytical shortcomings of any particular historians. It is rather to be found in the historicist template of social being which is used by all historians, in a model which obliges us to impose peculiarly modern metaphysical conditions upon the evidence of ancient experience. These conditions will always be amenable to similarly modernist, Eurocentrist readings of Athenian practices, just as they will always be distinctly uncongenial to the more historically sensitive alternatives proposed by the likes of Hunter, Foxhall, and Patterson.

For so long as we take it for granted that the Athenian *polis* was an essentially "political" entity, one where "power" radiated out from a central "democratic" control structure over other societal "fields," Athenian women will always already appear to us to be "excluded" from power and thus "dominated" by men. So long as we think of membership of this *polis* in modern-style, individualist terms as "citizenship," as a package of "rights" that entitled one to participate as an individual in the "political" or "public" realms, then it will always already seem to us like Athenian males systematically "denied" Athenian females their due "civic" privileges, treating them as something less than "equals" or "proper citizens," at best perhaps as *citoyennes passives*.

Likewise, so long as we think of the resources of *oikoi* in Attica in modern, economistic terms as "private property," then we will always already be inclined to see the male heads of those patrilineal households collectively as the self-interested, monopolistic "owners" of the wealth of the *polis*, while their wives will always seem to be little more than hired subordinates or exploited employees. Indeed, so long as we continue to presume that "humanity" itself can only be truly defined in modern individualist terms, as a kind of primordial personal entitlement to "life, liberty, and estate," the condition of the Athenian female will always appear to us to be one that is somewhat less than fully human.

Our standard analytical template thus all but requires us to reengineer non-modern experience at the ontological level, forcing it to conform to a peculiarly modern way of being human. It offers us no meaningful way to accommodate the possibility that women, as *politides*, were actually full and active members of the *polis*. On the contrary, it positively invites us to conclude that "misogyny" must have been "an important theme in Greek culture," that "Greek society" was essentially "patriarchal," and that ancient Greece offers us yet another chapter in the "history of the subordination of women." And in so doing, it inevitably prejudices the way we select and apprehend the "evidence" for gender relations in classical Athens.

Most obviously, the relatively small number of texts which appear to us to express a sweeping disdain or contempt for womankind come to assume an undue significance in our accounts. Thus, when we read, say, the Boeotian Hesiod's seventh-century story about how Pandora, the first woman, was sent by the gods as a "punishment" for man's theft of fire, or a light-hearted party song by Semonides of Amorgos (ca. 600 BC) which compares different kinds of wives to different species of animals, we tend to infer that such texts must speak authoritatively to the deep-seated misogyny of all ancient Greek males.[29] At the same time, when we encounter the far greater body of testimony which indicates, directly or indirectly, that women were routinely respected as complementary "partners" within the parameters of the time and routinely entrusted with essential life-sustaining responsibilities, we might be inclined to assign this other evidence somewhat less significance, to regard it with a certain suspicion, or even to see it as some kind of grand ideological smokescreen or subterfuge, since it challenges our own modernist presuppositions about the essential nature of "Greek society" as a dominion of male over female.

Needless to say, this is not for a moment to suggest that Athenian and Greek attitudes towards gender relations were worthy of modern admiration. We today are personally free to disdain their entire way of life if we so wish. But if we are to write meaningful histories of that way of life, we cannot dismiss their common sense "justifications" of it as "bogus," just because those justifications do not happen to align with our own uniquely modern presuppositions. We must instead try to make sense of that way of life recursively, on its own terms, in its own metaphysical conjuncture, as a distinctive way of being human that happened to be profoundly different from our own. And to appreciate the distinctively female contributions to that way of life on something more like their own terms, I suggest, we must at a minimum make the following adjustments to the way we objectify the *polis* and its prevailing gender regime.

Gender and Social Being in a Non-Modern World

First, and most fundamental, we need to dispense altogether with forms of social explanation which presume that Athenians, male and female, can be analyzed as if they were all natural, interchangeable, self-interested individuals. As our sources tell us again and again, male and female were designed expressly by the gods/nature to be different so that they could function as complementary "partners" in life. Since the various human skills, qualities, and dispositions necessary for life's production and maintenance were divided evenly between the two, neither one could subsist meaningfully in isolation as a free-standing being. Each one presupposed the other. Each one presupposed the mutually beneficial union of the two, a "coupling" (*zeugos*) that perpetuated the life of an *oikos*, just as each *oikos* presupposed a mutually beneficial "communion" (*koinonia*) with other *oikoi* that perpetuated the life of a *polis*.

Thus, to understand the basic logic of Athenian gender relations, one must effectively analyze husband and wife, in Foxhall's words, as a "single entity," as the male and female incarnations of a single, unitary *oikos* person or self. In other words, whenever the husband voted in assembly meetings, fought on the battlefield, or ploughed the fields, or whenever the wife nursed children, managed household finances, or made offerings to Athena, it was as if both were effectively participating in all of those activities simultaneously. So, strictly speaking, an Athenian could never act as a pure, unencumbered self, because one's very being always expressed the existence of one's *oikos*, which in turn expressed the existence of one's *polis*.

Second, we again need to objectify the *polis* as the *polis* objectified itself, as a symbiotic body of households, not as a functionally divisible societal terrain. Thus, instead of assuming that Athenian women were "excluded" from some entire "fields" of experience or "confined" to others, we would see that, as wives, mothers, workers, and ritual actors, they were self-evidently integral and essential to the life of the social body as a whole. Far from being continually "dominated" by any male-controlled "political" or "public" realm, they were continually expected and encouraged to exercise forms of responsibility, authority, control, discretion, and initiative in the management of almost all matters of everyday well-being, even if nature had also seen fit to design "partners" for them who could act on their behalf in certain specific locations, like assemblies, law courts, and battlefields. Athenian women were not simply the equivalent of modern "homemakers" or "housewives," beings who were left to tend to a "domestic sphere" while the real business of life went on somewhere else without them. In a world where the household itself *was* the real business of life, the many responsibilities that women routinely assumed in the conduct of that business were hardly modest, marginal, or insignificant. And in a world where households constituted the very cells of the social body, to contribute to the governance of an *oikos* was to contribute directly and substantially to the governance of the *polis* as a whole.

Third and last, we need to see in turn that modern categories like "democracy," "citizenship," and "civic rights" are less than useful in this particular historical context. As a form of *politeia, demokratia* encompassed the entire way of life of *polis*, not just a narrow "political regime." And as our sources insist, males and females were precisely equipped by gods to share responsibility for the way of life of every Greek *polis* as "partners." Being an Athenian was not about "citizenship" in any modern understanding of the term. It was not about at some point coming to possess a package of "civic" entitlements or "rights," an acquired relation to a "polity" that was external to oneself. It was about being born into a household that was already part of a very specific social body, a body that had been sustained and nourished since time immemorial by very specific gods and a very specific homeland. Under normal circumstances, an Athenian was not something one could "become." It was something one already was or was not, something one was naturally "made of," what one essentially was as a person. Clearly, there were two distinct and complementary types of Athenians, the male type (*polites*) and the female type (*politis*). Neither was somehow "more Athenian" than the other. Both were equally and necessarily integral members of the social body, even if the male type inevitably looks more like a "proper citizen" to us.

In sum, taking an ontological turn in our practice allows us to rethink the very foundations of social being in classical Athens, encouraging us to abandon inappropriate analytical models and unhelpful master categories, like the political and the social, public and private, democracy, citizenship, rights, and the natural presocial individual. In so doing, this alternative, recursive mode of analysis significantly changes the way we view the place of the *oikos* in the Athenian *politeia*, the numerous contributions made by women to that same *politeia*, and even the essential nature of the *polis* itself as an entity. And ultimately it helps us to see that life management in ancient Attica was categorically different from government in modern polities, because the primary producers of human order in the *polis* were its constituent households, the cells of the social body, not specialist administrative agencies or institutions.

That said, the business of reproducing the social body within the larger cosmic ecology of Attica could not be left entirely to each individual household. To continue our attempt to understand Athenian *demokratia* on its own ontological terms, we should look now at how *oikoi* were expected to work together as bodies or groups in the performance of various other ecologically essential functions, functions that individual households could not perform by themselves. In so doing, we will see further how the underlying principles of life management in ancient Attica differed categorically from those which underpin any modern mode of governance.

13

Living as One Liked

ACCORDING TO THE influential Foucauldian account discussed earlier, modern liberal government aims above all to produce and nurture a certain kind of human subject. Through its characteristic techniques and mechanisms of biopolitical rule, this government works on our minds, bodies, wills, and subjectivities to fashion each of us into a healthy, autonomous, self-directing individual, the historically novel kind of being that our science deems we must be if we are to thrive in our historically novel capitalist environment. Or, to put it another way, modern liberal government seeks to mould us into precisely the kinds of beings which modern liberal theory already presumes us to be by nature in the first place.

By contrast, the human subjects of Athenian government were not works in progress. They had no need for complex, professionalized mechanisms of education and life management, because they were already fully formed Athenians, already managing their lives individually and collectively in time-honored, ecologically productive ways.[1] For their *politeia* not only took for granted the existence of thousands of established, largely self-reproducing households. It also presumed that these households were already incorporated into various self-reproducing bodies and associations through commonalities of cult, blood, place, friendship, interest, and so forth. In other words, the Athenian way of life was not about developing new, ever more healthy, productive, and self-reliant forms of individual being. Among other things, it was about sustaining and perpetuating a complex armature of pregiven human interdependencies. It was about reproducing a primordial social body of social bodies, each one of whose members were naturally predisposed to collaborate with one another in mutually beneficial, life-enhancing group activities, performing vital ecological functions that individual households could not perform alone.

In this chapter we shall look at four such functions, all of them essential to the well-being of the *polis* ecology as a whole, which members of different households routinely and more or less spontaneously came together to perform for themselves.

Small Groups and Their Gods

The first and most urgent of these functions was again ritual in nature, namely the more or less continuous offering of prayers and gifts to shared patron gods, heroes, and ancestors all over Attica. As mentioned earlier, numinous agencies protected and supported a variety of smaller corporations in Attica, bodies which yoked together multiple households through a sense of common descent, friendship, and/or interest, such as phratries and *gene, orgeones* and *thiasoi, eranistai* and *theastai.*[2] At the same time, *oikoi* routinely convened to honor those superhuman powers who had a special concern for the inhabitants of particular localities within the *polis.* Venerable cult associations like the Tetrakomoi from the area around Piraeus and Phaleron and the Marathonian Tetrapolis in northeast Attica drew together constituents from several different settlements, even if the details of their practices remain somewhat obscure.[3] But far more significant were the rituals that were routinely performed in the 139 "demes" in Attica, each of which was at once incorporated within the totality of the "Demos of the Athenians" and an autonomous, corporate *demos* in its own right.[4]

As a communion of all the established households in a particular Attic village or district, each deme had its own calendar of sacrifices. Some of these events were essentially smaller, local versions of larger *polis*-wide rituals like the Theogamia, Skira, and Thesmophoria. Others were dedicated to the superhumans who were thought to govern life in that particular locality. Either way, these events were not reserved exclusively for male *demotai.* As with festivals celebrated by the *polis* as a whole, women played visible, sometimes important roles, both as priestesses and as general participants. Just as men naturally assumed all decision-making responsibilities both for their local *demos* and for the "Demos of the Athenians," so women routinely manifested their full and integral membership of both the local and the poliadic social bodies through the performance of life-sustaining rituals.[5]

To gain at least a preliminary sense of the scale and significance of this local ritual practice, we might look briefly at the case of the southern deme of Thorikos, where an unusually well-preserved calendar has been found.[6] The largely complete surviving portion of the calendar lists around sixty different sacrificial acts, some of which were held on the same days as components of larger ritual events. Roughly half of these sacrifices were for Olympian gods, including several each for Zeus, Poseidon, Apollo, Demeter, and Dionysus, and a small number of these were probably performed at sanctuaries outside the deme itself.[7] Other sacrifices were offered to numinous beings of more limited, local significance, such as Kephalos, Nisos, Philonis, "the heroines of Thorikos," and Thorikos himself, the deme's eponymous hero.[8] And here, as in other demes, it is also striking how local offerings were synchronized with those made by the entire *polis*. At Thorikos, as elsewhere, festivals and ceremonies like the Arrhephoria, Diasia, Plynteria, Pyanopsia, and Proerosia were evidently celebrated simultaneously by both Demos and at least some of its component demes, thereby ensuring a certain ritual consonance, and thus a kind of cosmic harmony, across the land of Attica as a whole.[9]

As far as one can tell from the remains of a small handful of other deme calendars, the ritual activities at Thorikos were not particularly unusual or exceptional. At a conservative estimate, each deme appears to have performed at least around forty sacrificial acts per year.[10] If so, this would mean that there were well over five thousand separate offerings made annually by demes alone, or roughly fourteen per day on average. When one then adds this figure to all those offerings made by individual households, by associations like phratries, *gene*, and *orgeones*, by bodies like the Marathonian Tetrapolis, and by the ten tribes, one soon begins to appreciate just how much the *polis* as a whole depended on all its constituent parts to perform the work of ritually reproducing the conditions of its existence.

Communal Legitimation of Marriages

The second ecologically vital function that groups of households were expected to perform more or less spontaneously was the recognition of marriages. There was no modern-style, centralized, legalistic system for certifying legitimate marital bonds between Athenians. But this did not mean that marriage in classical Athens was merely a form of "concubinage" or some kind of "common law" arrangement, whereby couples were essentially free to determine the nature and status of their own partnerships.[11]

As Cynthia Patterson in particular has stressed, marriage in Athens is perhaps best understood as a "social process." This process was of course initiated by the parents or guardians of the couple in question and then usually followed a highly specific sequence of socially acknowledged actions or events. As Patterson herself puts it:

> Although there were formal rules governing some aspects of the marital relationship, such as the laws on the heiress, marriage itself was recognized and validated not by a specific legal ceremony or piece of papyrus, but by the communally witnessed rituals and household events which over time established its legitimacy as an Athenian marriage. These included

the formal betrothal and designation of the bride as the future mother of legitimate children (*enguan*), the publicly witnessed wedding procession and feast preceding the wedding night itself (*gamein*), the setting up of a new household (*sunoikein*), and the birth and recognition of the children (*paidopoiein*).[12]

In other words, marital status in Athens was determined by common social knowledge and practice, not by any specialist governmental agency. One was legitimately married only in so far as one was continually seen and known to be so by one's relatives, friends, neighbors, and other acquaintances. And in this relatively stable, essentially endogamous, distinctly non-modern societal environment, where it could usually be taken for granted that subjects would actually have relatives, friends, neighbors, and other acquaintances on hand to see and know such things, this approach to the management of such an ecologically critical concern made perfectly good sense.[13]

By the same token, marriages in turn helped to reproduce these same vital interrelations, serving to build, extend, and fortify various kinds of bonds between households. Given the absence of any formal certification records, the evidence for particular marriages is inevitably rather piecemeal, coming mostly from items like gravestones and speeches in law courts. But enough survives to suggest some larger patterns.

Perhaps predictably, there was, for example, a general tendency towards intermarriage among neighbors, especially in rural districts, where residence and landholding remained most stable.[14] At the same time, there was a marked preference for marrying within an extended kin group, perhaps especially among wealthier Athenian families. In one particularly notable example, that of the distinguished Bouselid *genos*, as many as five different marriages between cousins are known to have taken place.[15] And if there was a certain self-evident logic at the time to pursuing marriages with neighbors, the idea of marrying relatives from one's extended family could also make very good sense. As Robin Osborne explains:

> Marriage is a bond that consolidates, it brings two groups into a strong moral link. The strategy employed in forming such a bond will depend on the particular circumstances, which themselves are not independent from the social status of the actors. The exchange of wives within the established kin group both reinforces links and can prevent the dispersion of property. In a situation where local bonds are strong and are maintained by other means, marrying kin, especially if kin are not local, may be important to prevent the disintegration of the kin group.[16]

Recognizing Athenians

The traditional strength of Attic bonds of locality and kinship also help to explain why phratries and demes assumed primary responsibility for our third critical ecological function, namely the recognition of eighteen-year-old males as full adult members of the social body of the *polis*, as beings who were thus capable of acting in the name of other Athenians.

Much about the phratries ("brotherhoods") remains rather mysterious to us.[17] But it is clear enough that they understood themselves and functioned as extended descent groups, incorporating male representatives from a number of specific, ostensibly consanguineous *oikoi*, who would convene regularly to perform common sacrifices for divinities like Zeus Phratrios and Athena Phratria. Under normal circumstances, induction into a phratry seems to have involved a very straightforward procedure, whereby a *phrater* would merely swear an oath that an inductee was his own legitimate son.[18] But in disputed cases, where other members questioned the inductee's legitimacy for any reason, more complex procedures could follow.

A good example is provided by an early fourth-century decree of the Demotionidai phratry, which appears to have had its roots in the Decelea area of northern Attica.[19] Here, in contested cases, it is prescribed that three members of the inductee's *thiasos* (apparently a sub-group of the phratry in this instance) must first give evidence under oath on the disputed issues at a preliminary hearing, then submit to a secret ballot on the candidacy, with the rest of the *phrateres* only voting thereafter if necessary. The matter is ultimately to be settled as follows:

> If the members of the *thiasos* vote that the candidate is a *phrater* of theirs, but the other *phrateres* vote him out, the members of the *thiasos* shall owe a hundred drachmas to Zeus Phratrios, except for any members of the *thiasos* who accuse him or are seen to be opposed to him in the adjudication. If the members of the *thiasos* vote him out but the introducer appeals to all and all decide that he is *phrater*, he shall be inscribed in the common registers. But if all vote him out, he shall owe a hundred drachmas sacred to Zeus Phratrios. If the members of the *thiasos* vote him out and he does not appeal to all, the adverse vote of the *thiasos* shall be valid. The members of the *thiasos* shall not cast a ballot with the other *phrateres* about children from their own *thiasos*. . . . The oath of the witnesses at the introduction of the children: "I witness that the one whom he is introducing is his own legitimate son by a wedded wife. This is true, by Zeus Phatrios. If my oath is good, may there be benefits for me, but if my oath is false, the opposite."[20]

The rather elaborate complexity of these procedures becomes readily understandable when one considers the stakes here for candidate, father, and phratry as a whole. After all, the votes cast in these cases had significant implications for the inheritance and thus the continuing life of an *oikos*, for the purity of the phratry's bloodstock, and of course for the very personhood of the candidate himself, as a true-born son, as an heir, as an Athenian *phrater*, and therefore, by definition, as a member of the Athenian social body. And for our purposes, what is perhaps most striking here is the freedom afforded to phratries to administer and determine matters of such momentous consequence for themselves. Apparently, since these bodies had a self-evident interest in preserving the purity of their own consanguinity, there simply was no need here for oversight by any other form of human authority.

That said, it still remains unclear whether all adult male Athenians were, by definition, members of phratries. But it is clear enough that the outcome of one's phratry candidacy case would have significantly affected one's fortunes when one subsequently came to be considered for recognition as a demesman. Since deme membership was likewise hereditary, and since all true Athenians necessarily belonged to *oikoi* that belonged to demes, to be recognized as a *phrater* necessarily meant that one was a legitimately born son of a demesman. Either way, the fullest ancient account we possess of the deme induction process, dating from the later fourth century, reads as follows:

> Those who are born from native parents on both sides (*ex amphoteron aston*) have a share (*metekhousi*) in the *politeia*. They are registered as members at age eighteen. When they are registered, the deme members take a vote about them under oath, first to decide whether they have reached the legally required age (and if not, the candidates return to the ranks of "boys"), and then to decide whether each one is a free man who has been born in accordance with the laws. . . . After this, the council (of 500) scrutinizes those who have been registered, and if anyone turns out to be under the age of eighteen, it punishes the deme members who have registered him.[21]

As this passage indicates, being an Athenian demesman was not a matter of "citizenship" in any modern sense, a status that was impersonally or institutionally bestowed upon one from without at the moment of one's majority. Under normal circumstances, like being a *phrater*, it was a form of personhood that one could only inherit from true-born Athenian parents, something that had already been decided at the moment of one's birth. So when it came to the crucial business of determining whether one was in fact such a person, whether one had in fact been born a full-blooded Athenian "according to the laws," who better to decide the issue than one's neighbors?[22]

In sum, marriage was hardly the only ecologically consequential "social process" in classical Athens. One might aptly apply the very same term to the means used by the Athenians to determine one's essential personality as a *phrater*, as a *demotes*, and thus as a future *oikos* head and parent of legitimate Athenian offspring. To be recognized as full and active members of the "Demos of the Athenians," male youths first had to be recognized by neighbors and kin as legitimate products and thus future reproducers of *oikoi*, demes, and phratries, the component parts from which the poliadic social body was made. To modern eyes, this way of administering such a vital communal concern might look like an elaborate "informal" or "pre-political" exercise, a kind of improvised, piecemeal response to the general absence of "strong" or "developed" specialist agencies of government. On the contrary, I suggest, it precisely exemplifies the Athenian approach to "government" in the broad sense, whereby it was just taken for granted that the component limbs, members, and cells of the social body would for the most part manage life's demands by themselves.

"Policing" Attica

To substantiate the point further, we need only consider the Athenians' general approach to our fourth essential ecological concern, namely the management and maintenance of human order in the *polis*, which likewise presupposed strong bonds between *oikoi*. In the only extensive modern treatment of "policing" in Athens that we so far possess, Virginia Hunter identifies two distinct if overlapping ways in which this ongoing communal self-regulation actually worked in practice.

The first of these she calls "gossip," which roughly corresponds to the Greek *pheme* ("what people say"), a force that proverbially "roams about throughout the *polis* of its own accord, broadcasting to everyone the details of one's personal conduct."[23] Drawing on various anthropological studies, Hunter observes that gossip effectively encourages conformity to shared norms by

> sanctioning individual conduct and thereby ensuring appropriate standards of community behavior. . . . For to criticize others is to imply some ideal standard of behavior from which one has veered. This observation holds for Athens, where gossip reached deep into individual lives. . . . Its circulation through a grapevine of neighbors and demesmen to local haunts, to village squares, and to the Agora in Athens indicates that no one

was immune from its criticism or ridicule. . . . To talk of others, whether idly or maliciously, was a deeply ingrained part of community behavior.[24]

As Hunter demonstrates, this ongoing "circulation" of gossip was perpetuated by all kinds of persons in Athens, male and female, Athenian and non-Athenian, free and unfree alike. And the most common subjects of this gossip can be taken as a fairly definitive guide to how members of the social body typically evaluated one another as Athenians. These included: the legitimacy of one's claims to being a true member of the *polis*; one's record of service to the *polis* in peace and war; one's treatment of one's parents and relatives; one's handling of one's patrimony; one's criminal acts; one's sexual mores; one's personal conduct and character; and the conduct and character of one's close relatives and associates.[25]

Gossip was most visibly involved in self-regulation in situations where an Athenian's personal background and contributions to the community were explicitly examined by a body of his peers. This occurred, for example, at the formal preliminary "scrutinies" (*dokimasiai*) undergone by would-be archons, councilmen, and other poliadic actors.[26] And it also occurred of course in trials before judge panels, where custom recognized that information about the everyday character and conduct of litigants was not only admissible but positively valuable as material evidence.[27] But gossip could also serve the cause of order in other, less formalized ways, especially when it came to pursuing the common interests of households. To take just one well-known instance, the cuckolded Euphiletus, mentioned earlier, here describes how he first learned of his wife's infidelity with Eratosthenes, a demesman of Oe:

> After this . . . an old woman approached me. I learned later that she had been sent by another woman whom Eratosthenes had also been corrupting (*emoikheuen*). This latter woman, who was angry and felt wronged by him because he no longer visited her so often, was on the look-out to discover what the cause of this might be. So after keeping watch for me near my house, this old person approached me. "Euphiletus," she said, "please do not think that I am approaching you out of any meddlesomeness. For the man who is violating (*hubrizon*) your wife happens to be a personal enemy of ours. If you take the servant girl who goes to market and waits on you and torture her, you will learn everything. The man who is doing this, corrupting your wife and many others, is Eratosthenes of Oe. He is very skilled in such behavior."[28]

When viewed from the vantage point of modernity, where one is accustomed to residing among complete strangers, where the management of all the larger, more urgent societal concerns is entrusted mostly to faceless specialists, and where indifference to the "private lives" of others is often deemed a virtue, the constant circulation of gossip about others in Athens might look somewhat unseemly or unduly intrusive. But in an environment where the human "governors" were all self-regulating "civilians," where the human face of the polity itself was no more than a body of interdependent households, where there could therefore be no such thing as privacy in any absolute modern sense, all this gossip was not just natural and inevitable. To a point it was positively healthy and productive, a form of knowledge that was essential to preserving the vitality of the social body itself.

But of course "what people say" was not always sufficient in itself to ensure "appropriate standards of community behavior." As Hunter goes on to show, when these standards were seriously violated, Athenians were also expected to regulate each other's conduct more actively, using their own resources, the help of friends, neighbors, relatives, and poliadic actors, and a range of legally prescribed mechanisms. To get an initial sense of how this second, more active form of communal self-regulation worked in practice, we can start by returning again to the case of the unfortunate Euphiletus.

After telling how he learned of his wife's affair with the *moikhos* ("seducer," "household violator") Eratosthenes from the old woman, Euphiletus then goes on to describe how his female servant and various friends and neighbors helped him to catch Eratosthenes "in the act" (*ep' autophoroi*) and bring the matter to a just conclusion.

> I was asleep. Eratosthenes, sirs, entered the house and the servant girl roused me straight away to tell me that he was in the building. After telling her to watch the doorway, I came down and went out in silence. I called on various friends one by one and found that some were at home and some out of town. Taking as many of these with me as I could, I went on my way. After getting some torches from the nearest shop, we entered the house. The door was open as the servant girl had prepared it so. After pushing open the door of the bedroom, those of us who entered first saw Eratosthenes lying in bed by my wife, while those who entered afterwards saw him standing on the bed naked. And I, sirs, hit him and knocked him down. Then pulling both his hands behind his back and tying them, I asked him why he was violating my house by entering it. He admitted his guilt and then besought and implored me not to kill him but to accept a sum of money instead.[29]

But the cuckolded husband was not to be persuaded. As Euphiletus puts it, Eratosthenes "met with the fate which the laws prescribe for such offenses." And this fate was in fact death, though Euphiletus discretely refrains from telling the judges how precisely he despatched his wife's lover.

While the killing of *moikhoi* caught in the act was surely not a common or everyday event in classical Athens, Euphiletus' story nevertheless serves to highlight three salient features of the Athenians' distinctly non-modern approach to the administration of "justice."[30]

First, courts did not have a monopoly over the production of legally binding outcomes in Athens. For certain kinds of offense, all adult members of the social body were deemed fully capable of administering the laws, as it were, extra-judicially, determining guilt or innocence and, if necessary, assigning and enforcing the designated penalty. True, for the great majority of offenses, standard procedure required a formal hearing before a panel of judges or other officials, whether the case involved harm against a particular person (*dike*) or potentially against the person of the *polis* as a whole (e.g., *graphe, endeixis, phasis, eisangelia*). But laws, written and unwritten, encouraged all Athenians to kill those who were caught in the act of committing certain kinds of capital offenses.

A particularly sensational, if somewhat exceptional example of such killing would be the fate that apparently befell one Lycidas after the battle of Salamis in 480. According to Herodotus, this man made the grave mistake of recommending that Demos consider a Persian peace proposal when he was serving as a member of the council of 500, which had to meet at that time on the island of Salamis following the evacuation of Attica.

> As soon as they heard this, the Athenians in the council, along with those who were outside, became enraged. Surrounding Lycidas, they stoned him to death.... Following the uproar about Lycidas on Salamis, the Athenian women found out what had happened. After bidding one another to follow, they went to the house of Lycidas of their own accord and stoned to death his wife and then his children.[31]

Such acts of summary group "justice," where more or less spontaneous gatherings of Athenians effectively excised a somehow harmful or "polluted" *oikos* from the social body, were again hardly commonplace.[32] But there were also specific laws which expressly authorized Athenians to kill various kinds of serious offenders on the spot, including would-be *turannoi*, homicides found in forbidden areas, night thieves, and of course *moikhoi* who were caught *in flagrante delicto*.[33] Furthermore, while it was generally assumed that those accused of less serious offenses would voluntarily comply with the designated procedures and appear for hearings and trials of their own accord, Athenians were entitled to arrange for the arrest of serious offenders, the *kakourgoi* ("malefactors") whom

they caught in the act of committing capital crimes, such as thieves of more valuable items (*kleptai*), kidnappers (*andrapodistai*), and temple robbers (*hierosulai*).
In such cases, one could either apprehend the criminal oneself and deliver him
to the appropriate volunteer magistrate by the procedure of *apagoge* ("leading
away"), or one could simply lead that magistrate to the miscreant by the procedure of *ephegesis* ("leading to"). Either way, one's action could lead directly to the
summary execution of said criminal without trial.[34]

Of course, the lawfulness of all such actions could later be challenged in a court,
either by the alleged offender himself or by his family in the event of his death.
Indeed, this is precisely what happened in the case of the killing of Eratosthenes, a
story we know about only because his family subsequently accused Euphiletus of
using the servant girl to entrap their deceased relative, thereby allowing themselves
to claim that the killing was a wrongful act of murder.[35] Yet what is most striking
about all of the cases described above is the degree to which all Athenians were
expected to assume a continual responsibility, both individual and collective, for
enforcing the most important laws of the *polis*, laws which dealt with the gravest
threats to the very life and well-being of the social body and its constituent cells.

Second, and perhaps more fundamental, it was entirely taken for granted
that Athenians would initiate all forms of legal action for themselves, whether
or not the case would be heard by a panel of judges. To be sure, the assembled
Demos might authorize the council and/or specially appointed persons to deal
with extraordinary, emergency-like situations, granting them temporary powers
to pursue investigations, assemble witnesses, make arrests, and if necessary punish
offenders.[36] But in the normal course of things, Athenian groups or individuals
were expected to initiate and pursue all legal procedures.

If the procedure in question was a case of personal injury (*dike*), only the injured person or family could pursue the matter. As for *graphai*, the other most
common category of case, they could be pursued by any male Athenian "who
wants to" (*ho boulomenos*), since they dealt with threats to the health of the
corporate self of the *polis* as a whole.[37] But in both kinds of procedure, a non-
specialist prosecutor was expected to take all the steps necessary to secure a
successful, lawful outcome. These steps could often include: searching out the appropriate laws; soliciting witnesses; gathering information pertinent to the case;
approaching the appropriate archon to trigger the issue of a formal summons;
participation in a preliminary hearing (*anakrisis*) with the defendant and appointed arbitrators; prosecution of the case before a panel of judges; and the
actual enforcement of any legally or judicially prescribed penalty.[38] As noted,
the case might be facilitated at various points along the way by sundry poliadic
actors, such as archons, arbitrators, a panel of judges, and the Eleven.[39] But we
should remember these too were all merely untrained, temporary volunteers by

modern standards. None were professional legal experts, because there were no such beings in Athens in any strict sense. Indeed, the only forms of professional help that were normally available to the prosecutor and the defendant were the services of *logographoi*, men like Lysias who made money from composing court speeches for others.

In short, what matters most here for our purposes is that the very will to apprehend, prosecute, and punish a miscreant in Athens, even those accused of the most serious, *polis*-threatening crimes, usually came from one or more members of the general body of Athenians, not from a specialist agency or apparatus of any "state." For the most part, it was up to "civilians" to determine whether a legal action should be pursued and even whether an offence had been committed in the first place. As Adriaan Lanni has emphasized, legal practice in Athens was thus fundamentally discretionary.[40]

Given this discretionary approach to regulating conduct in the *polis*, there was of course always the likelihood that members of the social body would use the laws to pursue and secure their own personal ends. And to a point, the Athenians accepted such eventualities, putting measures in place to prevent only the most blatantly unjust or frivolous prosecutions.[41] As Aeschines once acknowledged, there was a certain truth to the common perception in Athens that "personal enmities" (*idiai ekhthrai*) very often "set the community's affairs on the straight and narrow" (*ton koinon epanorthousi*).[42] And such attitudes seem to be entirely consistent with the Athenian way of life in general. There was no "legal system" as such in Athens, in the sense of an autonomous, superordinate, modern-style "machine." In the end, there was just an ensemble of laws, procedures, and volunteer functionaries, a set of tools and mechanisms which Athenian households could use to order and govern themselves, individually and collectively, on a continual, everyday basis. Thus, far from being some crude, "pre-modern" exercise in "self-help," Euphiletus's "execution" of Eratosthenes in his own bedroom illustrates precisely how Athenian laws were meant to be applied and enforced in practice. As Euphiletus himself apparently said to Eratosthenes, when he rejected the latter's plea not to kill him:

> It is not I who will kill you, but the established law of the *polis* (*ho tes poleos nomos*), which, in your transgression, you valued less than your own pleasures, choosing rather to commit a terrible offense against my wife and my children than to obey the laws and be an orderly person (*kosmios*).[43]

Third and more general, Euphiletus's case also well illustrates how legal practice in Athens presupposed a particular kind of social body, one whose members already possessed not just the wills but also the interpersonal connections, the shared interests, the collaborative predispositions, and perhaps above all the knowledge necessary to secure acceptable legal outcomes in this particular environment. Euphiletus's action against Eratosthenes simply could not have happened without a common knowledge of laws regarding *moikheia*, without the help of the servant girl, without the "gossip" of the old woman informer, or without the support of his torch-bearing friends, whom he gathered to serve as his collaborators, his protectors on the fateful night itself, and his potential witnesses if the case subsequently came to court. And this enlistment of help, information, and support of all kinds from a diverse cast of groups and individuals was again entirely characteristic of the way legal actions were pursued in classical Athens.

Of course, when mounting a prosecution, one could receive a degree of assistance with specific tasks like arbitration, apprehension, and imprisonment from the pertinent poliadic actors, so long as one asked for it. But as a general rule, prosecutors were left to manage their cases by themselves, not least because it could be safely assumed that they would receive most of the help they needed from friends, relatives, neighbors, and sometimes even bystanders or slaves.[44] And one should perhaps stress that these various helpers did not have to be male. For even if Athenian women could not prosecute cases or defend themselves in court in person, they were nevertheless entirely capable of helping and influencing the husbands and sons who did act as litigants, as a number of scholars have conclusively shown.[45]

Predictably, our social science lacks the necessary analytical terms with which to categorize this mode of legal practice. Terms that are based on modern experience inevitably presuppose the existence of a discrete "realm" of "the law" and a certain level of specialist expertise, both of which seem to have been entirely foreign to classical Athens. At the same time, terms like "popular justice" and "self-help" merely evoke the presumed antithesis of a modern "legal system," suggesting a relatively anarchic, spontaneous, and almost vigilante-like approach to order production, which is ultimately again unhelpful. It would be far better, I suggest, to see Athenian legal practice on its own terms, as another exemplary manifestation of a particular way of life, of a *politeia* that left members of a self-managing social body "free" to "live as one liked," because it was again taken for granted that they already possessed the knowledge and the resources to administer most of their own affairs for themselves, both as groups and as individuals.

Tribal Bodies

Before closing the chapter, a few succinct final words should be said about the *phulai*, the ten so-called "tribes" of Attica, which further exemplify the basic organizational logics and principles that we see elsewhere in the *politeia*. The tribes too defy standard social scientific categories. There has been some modern debate about whether they were essentially "political" entities or whether they are better seen as components of Athenian "civil society," as ancient equivalents of modernity's myriad non-political clubs and social organizations. In fact, they were neither, since both possibilities would again presuppose a modern-style, functionally compartmentalized societal terrain that was entirely alien to Greek experience. Again, the *phulai* are altogether better seen on their own terms, as the largest self-managing corporations within the Athenian social body, whose form they all replicated in miniature, since each one was formed by combining deme corporations from the city, hinterland, and coastal areas, the three principal sub-regions of Attica.[46]

The tribes likewise performed a number of functions that were essential to the life and well-being of Demos and the *polis* ecology as a whole. Most immediately, the *phulai* were each expected to furnish contingents of fifty volunteers annually from among their members (*phuletai*) for the council of 500, the body which was responsible for serving as the continuous point of contact between Demos, its constituents, and the outside world, for identifying and compiling all the items of business that Demos would consider during its forty annual day-long assemblies, and for maintaining a general oversight over both the conduct of officials and the condition of the communal resources of the *polis*, like its buildings and its triremes.[47] At the same time, the tribes were also expected to furnish armed forces for defending the *polis* on land in times of conflict, each one providing a regiment of infantry, a squadron of cavalry, and, at least originally, one of the ten annually elected generals.[48] Elsewhere, tribal affiliations were implicated in the organization of various other *polis*-wide practices and institutions, such as the *ephebeia*, the annual funeral for the war dead, law court trials, various supervisory boards, and even assembly meetings.[49] No less important, the tribes performed highly visible roles at major *polis* festivals, including the provision of dithyrambic choruses for the City Dionysia and teams of "warriors" for the armed dancing, torch race, and boat race contests at the quadrennial Great Panathenia, the most important festival of them all.[50] And more generally, through their members' shared experience of performing tribal functions and the complex, pan-Attic constitution of the tribes themselves, which ensured the presence of city-dwellers in every one of them, the *phulai* helped to foster and reinforce a sense of belonging to and "sharing in" the life of the larger social body of the *polis*, even among Attica's furthest-flung inhabitants.[51]

Given, then, the degree to which the very life and integrity of the Athenian corporate self and its ecology depended on the *phulai*, it is striking that these bodies were again largely left to manage their own affairs and activities, with little or no direction "from above," from Demos itself. Like demes, phratries, and *gene*, they had a corporate life all their own, despite their very high-stakes responsibilities.

All of the tribes were served by a number of functionaries, such as secretaries, treasurers, and heralds. But at any given time, general responsibility for overseeing the affairs of each one was assigned to three volunteer *epimeletai* or "caretakers," one each from the city, coast, and hinterland *trittues*. In typical Athenian fashion, these officials served only annual terms and performed their duties as a group, not as individuals.[52] Among their more important duties were the representation of the tribe at *polis*-wide gatherings, the stewardship of tribal funds and expenditures, and the convening and supervision of tribal assemblies (*agorai*). The business conducted at these assemblies likely included: addressing financial concerns; appointing the tribe's annual allotment of councilmen (*bouleutai*); appointing the tribe's annual liturgists, its designated sponsors of triremes, choruses, gymnasia, communal banquets, and other ventures; appointing tribal representatives for several *polis* boards, like those of the ditchmakers (*taphropoioi*), wallbuilders (*teikhopoioi*), and shipbuilders (*trieropoioi*); and awarding honors to members who had served the tribe with distinction, usually by performing the kinds of appointed roles just mentioned. As for the funding for these latter honors and sundry other tribal outlays, it seems to have come from a variety of sources, including land leases, personal benefactions from wealthier *phuletai*, and fines imposed upon members for infractions of tribal rules.[53]

Of course, given their complex tripartite composition and their relatively recent introduction in 508 BC, these tribes still tend to look to the modern eye like highly contrived, unnatural, or artificial bodies, mere technocratic devices that allowed a central authority to harness and mobilize the resources and manpower of an entire region to serve the common good of the *polis*. But that does not seem to be how their members experienced them.

Like the *gene* ("clans"), the *phulai* were constituted and acted as corporations whose membership shared a common line of descent, each one taking its name from its respective "ancestral founder" or "progenitor" (*arkhegos, arkhegetes*). Moreover, these tribal "ancestors" were all unusually illustrious, each one of them being a "hero" of the *polis* from the distant past, including among their number the former Athenian kings Cecrops, Erechtheus, Aegeus, and Pandion, along with Acamas and Antiochus, respectively sons of Theseus and Heracles, and the Salaminian Ajax, one of the Greek champions during the Trojan War. As such, all of these figures had long-established shrines in Attica, most of them in or around the areas of the acropolis and agora, and these were indeed the sites where the members of each *phule* would come together for their assemblies and their sacrifices to their "eponymous heroes."[54]

So even if the evidence appears to highlight the novelty and artifice of the *phulai*, telling us that they were only instituted in the time of Cleisthenes, when the Delphic oracle chose ten Attic heroes to serve as their eponyms from a long list of possible candidates, this data probably does not speak to the reality of life in these groups in subsequent decades.[55] For the hereditary nature of these groups, their proud associations with antique shrines and heroes, and, above all, their full integration into the everyday rhythms and routines of the "ancestral" *politeia* would together have encouraged an experience of their almost timeless fixity, their continuity with a distant past, and their proven necessity to the very existence of the *polis*.

Hence, it is no surprise to find that the *phuletai* of particular tribes were referred to collectively as if they were literal descendants of their heroic "progenitors," using the appropriate patronymic forms.[56] Probably the most vivid illustration of this sense of common heroic descent comes in a passage in the Demosthenic funeral oration, where the performances of each tribal unit at the fateful battle of Chaeronea (338 BC) are represented as continuing the familial virtues displayed by the ten eponymous heroes and their "original" offspring. A representative passage reads as follows:

> Thus, all of the "sons of Erechtheus" (*Erekhtheidai*) knew that Erechtheus, who gave them their name, allowed his own daughters, known as the Hyakinthidai, to be killed so that our land might be saved, giving them over to a certain death. So they considered that it would have been shameful if a man who was of divine descent had done the ultimate to preserve the freedom of his fatherland, while they, as mortals, valued their own mortal bodies higher than an immortal glory. Nor were the "sons of Aegeus" (*Aigeidai*) ignorant of the fact that Theseus, the son of Aegeus, had first established the principle that all Athenians could address *polis* assemblies (*proton isegorian katastesamenon tei polei*). They preferred rather to die than to continue living among the Greeks, clinging to their lives, with this principle lost . . . .[57]

In sum, the tribes appeared to themselves to be just as natural, just as historical, and just as integral to the *politeia* as all the other various forms of association that we have looked at in this chapter.[58] As such, they too were ultimately self-organizing bodies through which members of multiple households collaborated to perform essential ecological functions, functions that helped perpetuate the life of the *polis* as a whole.

The larger aim of these last three chapters has been to begin to piece together a robust alternative to standard modern accounts of "democratic Athens," where the *politeia* is reduced to an essentially secular, narrowly political "regime" or "system of government," one that was actively and exclusively controlled by male "citizens" through "public" mechanisms like the assembly, council, and courts. As we have now seen, a very different picture of this *politeia* begins to emerge when we try to analyze it recursively, on its own ontological terms, in its own metaphysical conjuncture.

Given a world where gods ultimately controlled the ecological conditions which governed all of life's processes and outcomes, this *politeia* was necessarily structured and organized around ritual activities, activities that were performed more or less continually throughout Attica by all manner of groups and persons. In a world where males and females were expressly created by divinity to serve as complementary, mutually dependent "partners" in the task of reproducing social being, the Athenian "way of life" necessarily recognized that both were full members of the social body. In a world where that same social body was ultimately comprised of self-managing households, it was just expected that householders, both male and female, would contribute to the reproduction of social being by meeting the most basic everyday needs of their *oikoi*. And in a world where households were routinely connected by ties of blood, friendship, locality, and membership of bodies like demes, phratries, and tribes, it was likewise taken for granted that Athenians would more or less spontaneously collaborate with others to take care of many of the essential everyday needs of the *polis*.

All of which begs a question. If Athenian *demokratia* was not then an ancient equivalent of a modern secular "political system," but an all-inclusive "way of life" that was pursued continually throughout the *polis* by all its members, male and female, how might this change our view of, say, assembly meetings, council activities, and court trials? How are we to understand the activities of the corporate person of Demos, the unitary totality of all "the Athenians," when it was visibly embodied by assembled *politai*? This is a question which we must now directly address.

14

The Cares of a Corporate Self

IF THE ATHENIAN *politeia* was a way of managing and reproducing social being that continually involved more or less all the humans in Attica, are we then obliged to classify the *polis* of Athens as a "stateless society," as some have argued?[1] I do not think we are. Even if it were truly helpful to define Athens by something that was not there, by the absence of a certain modernity, "stateless society" is too freighted a term. It implies a minimal progress along a single, apparently universal path of stadial development, whereby societal complexity is "achieved" only when a free-standing agency of rule has somehow detached itself from the rest of social experience in the modern style. If nothing else, our survey of the Athenian *politeia* has so far shown that a people does not need to objectify its social being in modern, spatial terms, as a functionally divisible societal terrain, to sustain a remarkably complex lifeworld.

As we have seen, human efforts to reproduce social being in Attica were quite highly routinized, guided and shaped as they were by an extensive inventory of well-established practices, procedures, mechanisms, and rules, written and unwritten. Moreover, even if the responsibility for performing all the work of this reproduction was continually dispersed throughout the social body, Athenians were not free to make and apply rules as and when they wished. There was an authority in Athens to which all were ultimately answerable, an ultimate maker and enforcer of rules, an agency which cared for all Athenian lives. And this agency was nothing less than the unitary person of them all, the corporate self they called Demos, whose will was manifest in all decisions made in assembly meetings and courts. So should we then think of this Demos persona as a form of "state"?

Again, I do not think that we should. For this Demos was yet another phenomenon in Athenian experience which does not readily correspond to any of our standard social scientific categories. And in order to help us apprehend this phenomenon and its role in the *demokratia* to which it gave its name, we can specify three significant ways in which it differs from a state, at least as that term is conventionally used and understood today.[2]

The Quiddity of Demos

First, the two differ in their essential nature or quiddity as entities. As social objects, they are quite incommensurable in that they presuppose fundamentally different metaphysical conditions.

In mainstream modernist social theory, the state tends to be objectified as a kind of autonomous command structure, an agency with its own distinct mind, will, and interests which rules over its corresponding "society," as it were, from without or above. As such, it is an essentially material entity, one that is in the end reducible to an objectively observable ensemble of practices, institutions, and human individuals.[3] In our standard accounts of Athenian "democracy," Demos of course tends to be defined in similarly modernist terms, as simply a material aggregate of individual "citizens," as the sum of all those living, breathing male persons who were equally entitled by right to govern the *polis* at any given time. And in some of these accounts, this means that Demos can then be roughly equated with "the masses," since "poor" citizens always significantly outnumbered "rich" citizens and the will of the majority always prevailed in communal deliberations.[4]

But Demos was not objectified in these modern-style materialist and individualist terms by the Athenians themselves. Within its own particular metaphysical conjuncture, it was a kind of ageless human superorganism, a unitary corporate self that had subsisted continuously in Attica since time immemorial, or at least since the time that Theseus had first integrated all natives of the region into a single *polis*.[5] As we saw in Chapter 10, it was a person that was ontologically distinct from the particular flesh-and-blood Athenian persons who comprised it at any given time, by turns preceding, incorporating, and outliving them all.[6]

It was this Demos, this deathless corporate self, that was the active subject of Athenian *demokratia*. Thus, the prevailing rationality behind this *politeia* was not a legalistic, egalitarian calculus, one that aimed to empower every native-born male individual in Attica as a political equal, as a rights-bearing, self-determining "citizen." It was about the self-reproduction of a unitary social body, not about enforcing a political equality among that body's adult male members. In other words, Athenian *demokratia* was whatever Demos and its many interdependent components had to do to ensure the continued vitality of the *polis* in an extremely precarious non-modern environment. Accordingly, it encompassed all of the various contributions to that common vitality that were made by Demos itself as an agency, by its component households, demes, and tribes, and by all of the many other groups in Attica, formal and informal, that directly or indirectly worked to perpetuate social being as a whole. In short, *demokratia* was wherever Demos acted in and upon the world, whether as a poliadic corporate subject or as constituent parts thereof. As such, its animating principle was unity not equality.

Of course, for the totality of Demos to act in and upon the world as a unitary person, it had to assume visible, material form. This happened whenever its constituents, male or female, convened to act in its person, whether in assembly meetings, law courts, at festivals, or on the battlefield. When so gathered, they would in effect shed their personal selves as *idiotai*, or household members, and assume their other social personae, as *politai*, as generic, interchangeable embodiments or instantiations of Demos, of Athens itself. Hence, if there was a dividing line in the *polis* between a human rule-making agency and the subjects of its rule, it was not some cleavage inscribed in the very fabric of experience between a "public" realm of the "state" and a "private" realm of "society." It was rather a line that was inscribed within the "dividual" subjectivity of each Athenian, between a *polites* personality and an *idiotes* personality, a line that allowed that very same Athenian to be in different circumstances both a particular member of a particular *oikos* and a generic member of the unitary Demos itself, to be effectively two different persons at once.

In order to find a social object that is even minimally comparable to the Athenian Demos in modern political thought, one would have to go all the way back to Hobbes's *Leviathan*. In that text, Hobbes too offers us a vision of a world which has yet to be dichotomized by any state-society divide, a world where an entire population of persons can be said to act as a single person, as a "real unity of them all," when it is physically instantiated and represented by some similarly unitary body or figure. But even here there are fundamental differences. As Hobbes himself makes expressly clear, his corporate person of the "commonwealth" or "Leviathan" is in the end just a *persona ficta*, a kind of legalistic fiction or conceit, since it exists only in and through the very act of its "personation" or representation by a "sovereign" figure. For in Hobbes's mechanical, materialist universe, where the only "natural persons" are material individuals, there can of course be no such literal thing as a corporate person.[7] In Athens, by contrast, Demos was a natural corporate person, a perpetually existing social body, one that did not have to be "personated" to exist. To the modern eye, it may of course appear to be no more than a conceit or a convenient fiction, a mere figure of speech. But when seen on its own terms in its own metaphysical conjuncture, where modernity's materialist and individualist truths simply did not apply, this Demos was self-evidently a "real unity of them all," the living human essence of the *polis* itself.[8]

Viewing Demos and *demokratia* on their own ontological terms in this way in turn changes the way we make sense of other characteristic features of the Athenian *politeia*.

For example, it helps us to see afresh the typical emphasis in Athens on inclusion and voluntarism in group decision-making, along with the widespread use of devices like lottery, stipends, and short term limits when it came to appointing all those officials and functionaries who served the will of Demos. Again, contrary to conventional wisdom, such practices were not shaped by any proto-liberal, individualist, egalitarian impulse or ideology. Rather, they helped to ensure the autonomy of Demos as an agency, as a corporate person that was visibly distinct from any of the particular persons who happened to embody it at any given time. For by continually dispersing administrative responsibilities among a sample of Athenians that was at once extremely large, more or less randomly chosen, and perpetually in flux, these practices helped to reproduce a reality where Attica was managed ultimately by a body of interchangeable *politai*, all of them generic incarnations of the *polis* itself, not by an aggregate of specific, nameable *idiotai*.

One might add that this essential distinction between *polites* and *idiotes* personalities was also reinforced by certain laws, which mandated stricter penalties for harming the former than the latter. The logic here is well expressed by Demosthenes in his speech *Against Meidias* (ca. 347/6 BC). As we learn in this text if an Athenian was somehow outraged or insulted as an *idiotes*, he had to prosecute the offender himself through *dike* or *graphe* procedures, which specified no particular punishment if the case were successful. But if that same Athenian was outraged or insulted as a *polites*, while performing his duties as, say, a *thesmothetes*, one of the nine archons, the offender would be punished with complete *atimia*, thereby effectively forfeiting his own capacity to act as a *polites*. For the offender in this latter case committed *hubris* not only against a particular individual but also

> against the laws, against the [thesmothete's] crown of office that belongs to all of you in common, and against the title [of that office] that belongs to the *polis*. For the title *thesmothetes* does not belong to any individual; it is a possession of the *polis* as a whole.[9]

In short, to harm a single poliadic actor or *polites* during the performance of his duties amounted to harming the whole Demos, the human essence of the *polis* itself.

At the same time, viewing Demos and *demokratia* on their own ontological terms also helps us to explain why Athenian records of assembly resolutions do not tell us who or how many attended or, for that matter, how many even voted in favor of the final resolutions that were produced. True, they do often tell us the names of the *rhetores* who proposed particular motions, because when one "advised" Demos one was an acting in one's own person as an *idiotes* and thus personally liable if the "advice" turned out to be somehow problematic.[10] But otherwise, they tell us simply that "Demos decided" (*edoxe toi demoi*) the issue. In the Athenian mind, the number and identities of the particular persons who actually voted was immaterial, because again the voters were not voting in their own persons as *idiotai*. They were voting as *politai*, as generic incarnations of the corporate, poliadic self, as "the Demos of the Athenians." Hence, to all intents and purposes, it was as if *all* Athenians had been there in the assembly, all voting as one for every resolution, with a single mind and a single will. Again, the animating principle of *demokratia* was unity not equality. The point of an assembly was not to serve as a venue where each Athenian male might express an equal right or power to pursue his own personal interest as an individual. The point of an assembly was to formulate and express the unitary, corporate interest of Demos, the human personality of the *polis* itself.

Likewise, seeing Demos and *demokratia* on their own ontological terms helps to explain the common practice in Athenian law courts of addressing panels of judges as if they were also responsible for all past Athenian court verdicts, all Athenian assembly resolutions, all Athenian military campaigns, all Athenian festivals, and so forth.[11] Because when litigants addressed panels of judges, they did not usually objectify them as a collection of discrete individual *idiotai*, but as *politai*, as the visible face of an autonomous Demos, the perpetual corporate self of Athens, which was of course ultimately responsible for all of the human actions and decisions that had been made by the *polis* as a whole to that point in time. At least in principle, the case was not being decided by a random assortment of flesh-and-blood individual persons. It was being decided by a single, deathless person, the transcendent person of Athens, whose life was the very history of the *polis* itself.[12]

So too, when texts speak of how Demos was temporarily "exiled" when the Peisistratid *turannoi* (ca. 546–510 BC) and the Thirty (404–403 BC) held power, they are not making far-fetched claims that all "supporters of democracy," never mind all flesh-and-blood "citizens" of Athens, were physically banished *en masse* from the city during the periods in question. Evidently, it came to be believed that individual opponents of *turannoi*, like Cleisthenes, Thrasybulus, and their respective allies, effectively assumed the personality of Demos, taking that persona with them when they themselves withdrew from the *polis* into exile.[13] And in the years following the restoration of *demokratia* in 403, it thus became common practice to address judges as if they themselves had all been collectively "in exile" with Thrasybulus and the other "men from Phyle," the men who had mustered in the Attic periphery expressly to resist the Thirty.[14] Of course, the personal, *idiotes* self of most if not all of these judges would have been physically present in Athens throughout the time of the oligarchy. But their poliadic self, the corporate self of the Demos that they now embodied as judges, had been elsewhere. According to conventional wisdom of the era, it had left the city when the Thirty took power. Their dividual selfhood thus allowed them all in effect to be in two different places at once.

And hence finally, when classical Athenian texts speak of tyranny, oligarchy, treachery, and other phenomena that would overturn or subvert the existing order of the *polis*, they consistently describe this process as being nothing less than a *katalusis* of Demos, a complete dissolution, decomposition, or disintegration of the corporate self of the *polis*, and thus of its *politeia*.[15] In standard modern accounts, this idea of "disintegration" makes little sense, because it is presumed that Demos is no more than an aggregate of atomized, material individuals in the first place. Hence the *katalusis* process is usually glossed in abstract, narrowly institutional terms, as the "subversion" or "overthrow" of "democracy," a mere change of political regime or form of government.[16] But if we see Demos and *demokratia* on their own ontological terms, we can begin to appreciate the full magnitude of what a *katalusis* meant to the Athenians.

It meant a lot more than a mere change in a form of constitution or government. It meant a lot more than the loss of any narrowly political rights or privileges for Athenian males. Potentially, it meant not only the dismantling of all the political, legal, religious, military, and other mechanisms that were required to sustain the life of Demos, the poliadic self. It meant the complete disaggregation and thus annihilation of this same corporate subject.[17] It meant nothing less than the death of the social body of "the Athenians," the extinction of the living human essence of the *polis*, and thus possibly the end of an entire cosmic ecology of gods, land, and people that had prevailed in Attica since time immemorial. Which is to say, it meant a change that was not just institutional but ontological, a change in the very fabrics of Athenian reality.

It thus goes without saying that the highest duty of any Athenian was to prevent any such ecological disintegration. It goes without saying that those, like traitors and would-be *turannoi*, who might cause such a catastrophic event, deserved only the ultimate punishment. For the stakes could not have been higher. And this helps to put into perspective cases like the prosecution of Leocrates by Lycurgus in 330 BC, mentioned earlier, which might otherwise seem to modern eyes like an exercise in juridical bullying or pettiness.

For Lycurgus, a highly influential *rhetor* of the time, it did not matter that Leocrates was a relatively obscure individual, a man who would otherwise be unknown to posterity. It did not matter that his offense was "merely" to leave Athens to pursue his own personal interests in Rhodes in the late summer of 338 BC, thereby violating an emergency order that required all Athenians to stay and defend the *polis* in a time of great peril, in the aftermath of the decisive defeat by

Philip II at Chaeronea. And it did not matter that the prosecution was taking place fully eight years later, in a time of less immediate peril, following Leocrates's eventual return to the city. He was a traitor to Athens, a man whose actions could have helped precipitate the *katalusis* of Demos,[18] thereby endangering the divinely sustained ancestral ecology of the *polis*. So, as the prosecutor vividly reminds the judges, it was not only the human constituents of Athens who would demand that Leocrates receive the ultimate penalty:

> If you acquit Leocrates, you will be voting for the betrayal of the *polis*, along with all its sanctuaries (*hiera*) and ships. If you have him executed you will be encouraging all Athenians to protect and preserve the land of our ancestors (*patrida*), along with all its sources of livelihood and well-being (*tas prosodous kai ten eudaimonian*). You need to recognize that the land and its trees are imploring you to help them, that the harbors, dockyards, and walls of the *polis* are begging you to help them, and that even the temples and the sanctuaries think it right that you should help them. So bearing in mind the charges against him, make an example of Leocrates, showing that the preservation of our laws and our Demos overrides any tears or pity.[19]

In the event, we learn, Leocrates was acquitted by the narrowest of margins, when the votes of the judges turned out to be evenly divided.[20] But texts like *Against Leocrates* start to take on a whole new valence and resonance when we take an ontological turn in our practice and view Demos and its *demokratia* recursively, in their own metaphysical conjuncture. Demos was not a modern-style "state," an aggregate of material individuals that ruled over an "Athenian society" from without. It was a continually self-ruling social body, the human personality of the *polis* itself, a body that defies our modernist standards of truth and realness. When Demos acted in or upon the world as a unitary corporate agency, all "Athens" was acting for itself, with a single mind, will, and purpose.

The Agency of Demos

This brings us to the second essential distinction between Demos and a modern "state," conventionally defined. While both are entirely capable of purposive action, the parameters and modes of their agency are quite radically different.

At least according to a certain modern mythology, the state is a proverbial "cold monster," a faceless, panoptic "machine" of domination, an agency that continually scrutinizes, polices, disciplines, and corrects the deeds, the words, and even the thoughts of its subjects.[21] Whether one sees it as, say, a neutral bearer of a traditional juridico-coercive sovereignty, as the organizing committee of the bourgeoisie, as a general patriarch, or merely as the designated coordinator of knowledge-power for a prevailing, impersonal truth regime, one can scarcely doubt that this agency plays an active, interventionist role in ordering the very conditions of modern existence. One simply cannot question the extraordinary capacity of what Michael Mann has called the "infrastructural powers" of the state "to penetrate civil society" and "to implement logistically political decisions throughout the realm." As Mann goes on to explain:

> These [infrastructural] powers are now immense. The state can assess tax on our income and wealth at source, without our consent or that of our neighbours or kin (which states before about 1850 were never able to do); it stores and can recall immediately massive amounts of information about all of us; it can enforce its will within the day almost anywhere in its domains; its influence on the overall economy is enormous; it even directly provides the subsistence of most of us (in state employment, in pensions, in family allowances, etc.). The state penetrates everyday life more than did any historical state. Its infrastructural power has increased enormously. [F]rom Alaska to Florida, from the Shetlands to Cornwall there is no hiding place from the infrastructural reach of the modern state.[22]

When compared to this restless, activist state of modern experience and imagination, the Athenian Demos seems strangely passive, even inert as an agency. Far from being any kind of panoptic, interventionist automaton, its intentionality, its sense of a need to act, its very will to act, came from elsewhere. And far from exercising an "infrastructural power" that allowed it to "penetrate" the fabrics of everyday life, it mostly just authorized others to act in its name. For in the end, when materialized as an agency in assembly meetings and courts, Demos per se was just a deliberative rule-making body. It did not actively and continually "govern" Athens at all in any modern sense. Its primary role in the *politeia* was simply to respond to communal problems and issues by producing binding, authoritative decisions, decisions which helped fix and maintain the conditions under which all Athenians throughout Attica, as both *idiotai* and *politai*, would continually govern themselves.

Thus, for example, the list of motions (*probouleumata*) to be considered by Demos at all assembly meetings had to be prepared by another body, the council of 500. And the agenda items themselves could be initially suggested to the council by a wide range of different persons, including the council's own members, any Athenian *idiotes*, and representatives from other *poleis*. Moreover, during *ekklesia* meetings, it was up to particular "advisors," speaking in their own persons as *idiotai*, to frame and propose specific courses of action on the matters under consideration, since Demos, as a unitary body of the whole, could not advise or deliberate with itself. It could only decide between the alternatives presented, with final resolutions (*psephismata* and *nomoi*) determined simply by a majority vote. Likewise, it did not itself pursue legal actions when materially embodied as a panel of judges, since every routine law case in Athens had to be initiated and prosecuted by an *idiotes*. Demos was responsible only for resolving the issue by determining guilt or innocence and, if necessary, for choosing between the punishments proposed by each side.

Moreover, this same rule-making Demos could not even directly enforce its own decisions, whether they took the form of assembly resolutions or judicial verdicts. Strictly speaking, all it could do was to empower others to realize its will. In practice, this usually meant delegating responsibility for enacting specific decisions to generals, treasurers, boards of officials, and other poliadic actors, then depending on the council of 500 to oversee, coordinate, and ensure the successful performance of the assignments in question.[23]

To illustrate further the remarkable limitations of Demos as a rule-making agency and get a more precise sense of how it functioned in practice, we might briefly consider the trial and execution of Socrates, a prominent critic of *demokratia*, in 399 BC. Many moderns, of course, have wondered why "the Athenian state" saw fit to execute one of Greek antiquity's most celebrated intellectuals, thereby rendering him something like a "martyr" to the cause of free thought and free speech. To all appearances, as I. F. Stone once observed, "[w]hen Athens prosecuted Socrates it was untrue to itself."[24] But this martyrdom account misapprehends the agency of Demos and the business of Athenian rule-making in at least four fundamental ways.

First and most important, "Athens" did not prosecute Socrates. Like all routine legal procedures in Athens, the entire case against the philosopher was prepared, initiated, and prosecuted to its conclusion by an *idiotes*, namely Meletus of Pitthos, who was supported by two "co-speakers" (*sunegoroi*). By following the procedure for a *graphe asebeias* ("indictment for inappropriate engagement with gods"), these men claimed to be pursuing a matter that affected the well-being of the *polis* as a whole. But at no point in the proceedings did they themselves act as or for Demos, as *politai*. In short, none of the intentionality behind the case, none of the original will to punish Socrates, came from the human personality of "Athens" at all.

Second, throughout the entire process, all those who *did* act as and for Demos, as human incarnations of "Athens," retained the stance of a strictly neutral third party, one that was simply trying to facilitate the resolution of a dispute between two *idiotai* so as to produce a communally acceptable outcome. As an ensemble of actors, they were all just non-specialist volunteers, performing their duties for a single year. They did not in any way resemble an autonomous state-like command center, one that was capable of, say, targeting and persecuting dissident intellectuals in the name of some compelling, self-interested *raison d'état*.

Thus, the *archon basileus* was required only to convene the preliminary *anakrisis* ("examination"), where he merely established for the record the basic details of the charge, the primary evidence, and the plea of the defendant, then set a date for the trial. He neither investigated the alleged crime nor evaluated the merits of the case in any rigorous fashion.[25] Likewise, the Eleven, who administered the jail and oversaw executions, did little more than follow the directions prescribed by the pertinent legal procedures and judicial verdicts. Nor were they personally responsible for killing Socrates or any others who were condemned on capital charges.[26] As for the five hundred judges in the case, the judicial face of Demos itself, they were required only to listen to the timed speeches of Socrates and his opponents, to determine his acquittal or conviction by a single majority vote, and then to choose between the alternative punishments suggested by the two sides. Their duties conspicuously did not include challenging the claims of litigants, cross-examining witnesses, or questioning the validity of evidence. In fact, at no point during the trial were they obliged or expected to speak at all. Their decisions to convict and to execute Socrates were thus reached without any formal intervention in the trial proceedings, without any deliberation among themselves, and without any expert legal guidance of any kind.

Third, Socrates was not prosecuted by Meletus for his non-conformist ideas, since Demos did not criminalize heterodox thought or speech per se, however potentially subversive. Apparently, he was prosecuted for two specific actions, namely "introducing other new kinds of deities" and "corrupting the young."[27] The former charge referred presumably to the personal *daimonion* ("spirit") which, as Socrates was apt to claim, guided his everyday conduct. The latter was a coded reference to the known fact that some of his youthful followers had later gone on to become members and supporters of the Thirty, not least Critias, the notorious figurehead of that regime.[28] Be that as it may, these were extremely grave charges. As this study has made clear, to threaten the integrity of the Athenian pantheon and the vitality of Demos itself was to endanger the very existence of the *polis* ecology upon which the lives of all depended. Of course, one is free to debate the fairness of such charges in this particular instance.[29] But it is indisputable that Socrates harmed his own cause by refusing to treat the accusations with the seriousness that they deserved. And given the philosopher's general reputation as an opponent of the established *politeia*, along with all the recent threats to the very existence of that same *politeia*, it is not at all hard to see why the judges might have decided to resolve the dispute in question by authorizing his execution.[30]

Fourth and last, Socrates' treatment did not at all violate communal norms regarding freedom. Indeed, his case exemplifies the extraordinary threshold of freedom that was granted to defendants in all conventional Athenian legal processes. As far as one can tell, he was not physically apprehended by any poliadic actor at any point prior to his incarceration. And during the entire time from his original indictment to his final condemnation, he was free to exercise at least some degree of choice and discretion. Thus, at the preliminary "examination" with the king archon, he was free to challenge the charges against him and free to question his accuser. Later, when on trial, he was of course free to persuade the judges of his innocence and, failing that, free to propose a suitable punishment for himself. And at almost every point before, during, and even after his trial he was effectively free to escape into exile from Athens with little fear of pursuit. Famously, he chose to remain and face his punishment.[31] But even then, as his final hours approached, a certain margin of freedom remained. For he was still free to receive visitors, free to choose those in whose company he would spend his final instant of life, and free even to choose the precise moment at which to drink the hemlock, to choose the exact time of his demise.[32]

In sum, when seen on its own terms, in its own metaphysical conjuncture, the Demos of the Athenians was all but incapable of producing "martyrs" to the causes of free speech or free thought. As the ultimate rule-making agency and human essence of the *polis*, as the very personality of "Athens," it remained entirely "true to itself" throughout the prosecution of Socrates. Far from suppressing the philosopher's "democratic" liberties, Demos and its various auxiliaries behaved in a manner wholly consistent with *demokratia* as the Athenians knew it, as a way of life where all members of social body, even dissident intellectuals, were free "to live as one likes" within the limits of communally accepted laws and customs.[33]

The Competence of Demos

The third and final essential difference between the Athenian Demos and the modern master category of "the state" requires only brief discussion. For in any such comparison, one can hardly avoid noticing the remarkably limited functional competence of Demos as an agency. If its primary responsibility in the *politeia* was just to produce binding, authoritative rules, the rules in question were for the most part restricted to a rather narrow and quite specific range of matters. Whether these rules were framed as laws (*nomoi*), assembly resolutions (*psephismata*), judicial verdicts (*kriseis*), or elections (*hairesiai*) or reviews (*dokimasiai*) of officials, they concerned only matters which affected the life of the social body as a whole.[34]

As we saw in the previous chapter, it was just taken for granted that households and groups of households would assume responsibility for managing their essential everyday life concerns, including routine ritual performances, health, education, marriages, "police" work, and of course the material well-being of the households themselves, the cells of the social body. Demos did not even monopolize rule-making in Attica. As we also saw earlier, bodies like demes, tribes, and phratries were largely free to make and enforce their own rules and decisions on ecologically critical matters, like the management of cults of gods and heroes, military levies, and admissions to the Athenian social body itself, so long as they did not contravene the established laws and customs of the land. As a general principle, in all of this ongoing daily business of life management, Demos would involve itself only when it was invited to resolve disputes that arose between parties.

Otherwise, the competence or reach of Demos as an agency was generally limited to producing the kind of *polis*-wide rules which the many component parts of the social body could not be expected to produce and administer for themselves. As one might anticipate from our discussion so far, these rules covered three general areas of ecological concern: the *hiera* of the *polis*; the *hosia* that were held in common by all members of Demos; and relations with other peoples in the world beyond.[35]

The first and most important of these areas of course concerned the *hiera* of the *polis* and the ongoing relations between Demos and its gods. Custom required that *ta hiera patria* ("the ancestral rituals") were always the first matters to be considered at every assembly of Demos and every council meeting.[36] And at least three such matters had to be discussed at every other session on the Pnyx.[37] For when it came to the management of relations between "the People of the Athenians" and any divinities, Demos itself was necessarily the ultimate source of all binding decisions. As Robert Parker has observed:

The people decides what gods are to be worshipped by what rituals at what times and places and at what expense: it regulates too the duties and terms of office of priests and priestesses, and creates new priesthoods at need. Persons claiming special expertise are free to contribute to debate and to suggest new measures, and may be invited to bring special topics to the assembly. Their voice will often be listened to with respect, but they have no power to enforce what they advise. On many such topics the assembly decided to seek the guidance of an oracle, and a clear oracular answer on a matter of cult was always treated as binding. But a formal motion to accept the advice of the god was still required: and, more important, the decision to take the problem to the god was itself made by the assembly. Binding and enforceable religious rules emanated only from there.[38]

Responsibility for implementing these rules on the ground then fell to the council, whose duties included overseeing and, if necessary, appointing ritual officials, co-ordinating relations between officials and priests, keeping track of all relevant incomes and expenditures, especially those from the *temene* and treasuries of the gods, supervising the plans and contracts for cult buildings and monuments, managing the production of festival prizes and artifacts like the annual *peplos* for Athena, and so forth.[39]

If relations with gods were necessarily the first item of business in any assembly or council meeting, the second was relations with other peoples elsewhere, the business that was transacted through "heralds and embassies" (*kerukes kai presbeiai*).[40] Again, Demos had to authorize all diplomatic activities by Athenians, whether this meant selecting and instructing ad hoc "ambassadors" (*presbeis*) and others to act abroad on its behalf or directly voting on matters like the clauses of treaties and armistices, ultimata from other *poleis*, arrangements for grain imports from elsewhere, the appointment of non-Athenians as honorary agent-representatives (*proxenoi*) of the Athenians in other lands, and the award of honors to any non-Athenian "benefactors" (*euergetai*) for services rendered.[41] Likewise, if conflicts arose, only Demos could decide significant issues like war declarations, the size of forces, the appointment of commanders, the funding required, and the larger strategic objectives of campaigns. And again it was left to the council, coordinating its efforts with commanders and other poliadic actors, to oversee logistical details, like ensuring the availability of the necessary funds, manpower, and *matériel*, maintaining ships and dockyards in a suitable condition, commissioning the design and construction of new vessels when required, and monitoring the performance of trierarchs.[42]

The third and final area of concern for Demos was the administration of the *hosia* of the *polis*, things held in common by all Athenians.[43] Under this rubric, one might include, for example, the management of water supplies, roads, coins and measures, market-related and other common-use buildings and spaces, mining leases, harbor revenues, pension disbursals for invalids, resources forfeited by convicted criminals, and dung collection. Again, if significant issues arose concerning such matters, Demos itself would vote directly to resolve them. Otherwise, the day-to-day administration of these matters was again handled by the council of 500, again often acting in concert with other *polis* actors, such as the boards of sellers (*poletai*), marketplace managers (*agoranomoi*), inspectors (*epistatai*), and auditors (*logistai*).[44]

Self-Management by the Parts and the Whole

Pulling the findings of these last three chapters together, we can see how analyzing *demokratia* on its own ontological terms, in its own metaphysical conjuncture, significantly alters our perspective on human "government" in Attica. Perhaps above all, this approach offers a salutary counterpoint or corrective to the modern tendency to fixate on the activities of the corporate person of Demos, as if mechanisms like assemblies of householders, panels of untrained judges and councilmen, and lottery-filled magistracies somehow together constituted an ancient equivalent of modernity's specialist, machine-like "states."

As we have now seen, this corporate rule-making Demos did not actively and continually govern everyday life in Attica, "penetrating" an Athenian "society" with its despotic infrastructural powers. Along with its auxiliaries in the council and elsewhere, it just resolved matters in court cases and in those three general areas of life which Athenian males and females, as householders, neighbors, friends, relatives, and members of various associations, could not readily manage for themselves. Even then, within the distinctly limited parameters of its competence, the corporate person of Demos did little more than produce binding rules on issues that were raised by others, most obviously by all those *idiotai* who took it upon themselves to frame assembly proposals and prosecute court cases. And of course even then, when it had assumed material form and decided issues in courts or assembly meetings, Demos did not act as a free-standing, state-like agency, one that imposed its will upon a social body from above or without. It *was* the social body incarnate, a "we" not a "they," the visible human essence of the *polis* itself.

In other words, the larger point here is not that we need to replace our standard "top-down" model of Athenian order production with one that is instead "bottom-up." To understand *demokratia* in its own terms, we need to recognize that there really was no "top" or "bottom" in Athens, no rule "from above" or "from below." Instead, there was just a self-ordering social body in a particular ecological environment. And this body could manage itself either severally, as self-regulating, symbiotic "parts," or when necessary as the unitary totality of the "whole." To think of this entity as any kind of "society," whether state-centered, stateless, or something in between the two, is to misapprehend its essential nature.

If the Athenian *politeia* was thus an ongoing exercise in ecological self-management by a social organism and its many interdependent components, all of which helped to reproduce the life of the whole by reproducing themselves as parts, this new account of *demokratia* in turn alters our perspective on the production, distribution, and exchange of life-sustaining resources in Attica. In the following chapter, we will complete our survey of the *polis* ecology by considering the distinctly non-modern, non-capitalist logics which governed the circulation of vital resources, the continual traffic in the sources and means of life itself.

15

The Circulation of Life's Resources

ONE COULD FAIRLY question the remarkable weight that is placed upon the category "economy" in both the analysis and the management of modern societal well-being. One could reasonably point out, for example, how prevailing social knowledge conventionally reifies this category, treating a mere metrological device as if it were somehow a mind-independent, machine-like thing-in-itself, a self-regulating "system" of wealth allocation that seems to actively produce the very statistical averages and aggregates from which it has been fabricated. And one could surely observe that this device both presupposes and perpetuates an extraordinarily narrow definition of "wealth," effectively reducing national well-being to a monetarily measurable value. In so doing, it expressly excludes from its compass many other possible, less readily monetizable indices of vitality, not least the health of geophysical processes and non-human lifeforms, human minds and bodies, marital and familial relations, communal bonds and cultural attachments, standards of morality, justice, and equity, and general senses of happiness, fulfilment, and purpose.[1]

That said, the idea of a self-ordering "economic system" does at least make ontological sense as a phenomenon in a modern lifeworld, one that rests upon materialist, anthropocentrist, secularist, and individualist foundations. And to the extent that the prevailing structures of our capitalist way of life oblige us to act as if such systems are always already there in the fabrics of our experience, continually determining the material fortunes of one and all, economies clearly enjoy a contingent realness as such in modern experience. But we should not mistake this contingent modern realness for a universal objective realness. By their very nature as entities, economies cannot be realized in lifeworlds where the metaphysical conditions of their possibility are simply missing. People cannot act on the assumption that their fortunes are governed by economic systems when they cannot imagine the existence of such systems in the first place. Accordingly, when our mainstream historicism predisposes us to organize our accounts of non-modern worlds like classical Athens around the category "economy," to abstract from the evidence of experience circulations of resources which to modern eyes possess a narrowly economistic "value," the results will inevitably be problematic, as we saw earlier in Chapter 3.

To produce more ethically defensible, more philosophically robust, and more historically meaningful accounts of the flows of essential, life-sustaining resources in non-modern worlds, I suggest, we again need to analyze them recursively, using local, non-modern "models" rather than those of modern economics.[2] This again means asking how our historical subjects objectified the essential givens of their experience, how they made sense of themselves as persons, their accepted modes of sociality, and the physical and metaphysical environments in which they lived. More particularly, it means reconsidering what our subjects might themselves have deemed to be "resources" that were at once "essential" and "life-sustaining." It means understanding their prevailing notions of health, wealth, and happiness, their a priori conceptions of value, and ultimately their taken-for-granted ideas about the sources, means, and ends of life itself.

Turning then to the particular case of classical Athens, a recursive approach would clearly require us to trace those flows of resources which sustained the cosmic ecology of the *polis*, uniting gods, land, and social body in Attica into a single symbiotic system of life. It would also require us to explore the circulations of resources within and between the various components of the social body, such as households, demes, phatries, and tribes. And it would require us to recover the various resource exchanges that were transacted between those sundry parts of the social body and the unitary totality of Demos, the poliadic self.

Instead of trying to describe all of these ecology-sustaining resource flows in exhaustive detail, an undertaking which would require an entire book in itself, the following chapter will instead attempt to illustrate the value of a recursive approach by highlighting four fundamental ways in which it would further alter our perspective on the production of life and well-being in Attica.

Exchanges between Humans and Superhumans

First and foremost, a recursive approach would help us to see that the most essential "exchanges" in the *polis* were not those transacted between persons in markets, ports, and other commercial settings. Rather, the most essential exchanges in the *polis* were those that were transacted between the people of Athens and its gods, whereby tithes, sacrifices, libations, votives, temples, and other *hiera* were "traded" for vital resources such as sunshine, rainfall, agricultural fertility, bodily health, and effective decision-making. To a point, gifts to the gods were like taxes rendered to maintain the infrastructure of the cosmos. By continually transforming *hosia* into *hiera*, the Athenians secured the basic conditions of their own existence. And it would be hard to understate the sheer scale of these transactions or their larger significance for the *polis* ecology as a whole. A few examples will help to substantiate the point.

Consider first the extraordinary array of resources and energies that were invested in the construction of the Parthenon, the single most lavish gift that the Athenians ever dedicated to the divine custodian of their *polis*.[3] Realizing this *grand projet* required resource contributions of various kinds from at least a quarter of all adult Athenian males, along with dozens of metics and slaves.

In the first instance, the project of course required deliberations and votes by the thousands of assemblymen who had originally considered the plan and approved it in the early 440s. It required the more or less continuous attentions of thousands of councilmen, dozens of "commissioners" (*epistatai*), and the various other poliadic actors who subsequently monitored and supervised progress on the temple during the many years of its construction (447–432 BC).[4] It required physical labor by perhaps as many as a thousand workmen, from relatively unskilled quarrymen, carters, and builders to the more specialist craftsmen who executed the decorative details of the temple and its celebrated chryselephantine statue.[5] And it demanded the efforts of an extraordinary number of animals to transport the temple's massive column drums and other marble membra the eleven-mile distance from the Pentelikon quarry to the acropolis. According to one recent calculation, this task would have required on average as many as fifty-five pairs of oxen each day over a period of some ten years, with teams working primarily in late summer and early fall when the roads were most readily passable.[6]

All this is of course to say nothing of the colossal financial costs incurred by such a monumental undertaking. In the absence of detailed accounts, the exact figures involved cannot be known. But the most thorough and widely accepted modern estimate would reconstruct the costs of just the stonework for the Parthenon as follows:

1. Transport

	Tons	Cost in drachmas
Walls	7,835	88,144
Pavement	3,819	42,964
Colonnade	10,186	114,592
Core	8,941	43,811
Total	*30,781*	*289,511*
		= 48 talents

2. Quarrying, erecting, and polishing

	Tons	Cost in drachmas
Walls	7,835	484,203
Pavement	3,819	228,376
Colonnade	10,186	958,503
Core	8,941	113,327
Total	*30,781*	*1,899,922*
		= 317 talents[7]

When we then add on the various other expenses involved, like those for the roof, ceiling, woodwork, and sculpture, the overall cost of the project becomes clearer still:

	Cost (drachmas)	Cost (talents)
Stonework	1,899,922	365
Ceiling, roof, gates	390,000	65
Pediments, acroteria	105,800	17
Parthenon frieze	70,875	12
Metopes	60,000	10
Total	*2,526,597*	*469*[8]

So how exactly are we to translate these figures into intelligible modern values? Conventional wisdom tells us that the average wage for a skilled manual worker, like a good number of those who would have worked on the Parthenon, was one drachma per day, so very roughly an equivalent of, say, US $100 ($12.50/hour for an eight-hour day).[9] If the total cost of the building was somewhere just over 2,500,000 drachmas (469 talents), this would make the modern U.S. equivalent cost around $250 million, an extravagant sum in any age.

One might add that the scholar who originally came up with the 469 talent figure readily concedes that it may be "surprisingly, if not impossibly, low," with the real figure perhaps being "well over 500 talents."[10] We should also note that this enormous figure would have been significantly higher if it had included the costs of the marble itself, the building's primary fabric. But since the quarry on Mt. Pentelikon was considered to belong already to the Athenian Demos, as a gift from the gods, these raw materials actually cost nothing at all. And we should note finally that these figures also do not include the costs of the building's famous statue of Athena Parthenos. By some reckonings, this extremely lavish effigy, with all its gold and ivory decoration, may have cost as many as 850 talents in itself, bringing the total cost of the temple up to 1,319 talents, or very roughly US $790 million.[11]

To put this figure into some further perspective, it was around triple the total sum of annual "contributions" which the Athenians typically extracted from over 150 subject *poleis* when their empire was at the peak of its strength in the 440s and 430s.[12] It was also a figure sufficient to pay for the continuous manning of a substantial naval force of 100 triremes for over a year.[13] Indeed, it would have covered most if not all of the costs of the nine-month blockade and siege of rebellious forces on the allied island of Samos in 440, an unusually expensive, demanding campaign that at times required a hundred or more Athenian warships to stifle a serious threat to the integrity of the empire.[14] Perhaps more to the point, the funds expended on building the Parthenon could have sustained an average-sized *polis* of 2,000 or so *oikoi* at or slightly above subsistence level for more than ten years.[15] And assuming a similar general standard of living, these same funds could have sustained a *megalopolis* of 40,000 *oikoi*, roughly the size of Athens itself at the height of its prosperity in the fifth century, for the better part of seven months.

If temples were typically among the most costly gifts that the Athenians could dedicate to their gods, these buildings also routinely served as storehouses for a variety of smaller *hiera*, votive "treasures" that expressed human gratitude for continuing divine support. This was especially true of temples on the Athenian acropolis, above all of the Parthenon itself. Recounting a speech given at the start of the Peloponnesian War by Pericles, who was anxious to reassure Demos that its collective resources dwarfed those of their opponents, Thucydides gives us an impression of the enormous quantities of the treasures that belonged to Athena and other gods in Athens at the time (ca. 431/0 BC):

> And he bade them take courage that, apart from any other revenue, the average annual contributions of the allies alone amounted to six hundred talents. And there was still on the acropolis a sum of six thousand talents of uncoined silver (left from the 9,700 talents that once been there, after money had been used for the acropolis gateway, for other buildings, and for [the siege of] Potidaea). Moreover, there was all the uncoined gold and silver in both personal and communal votives, the consecrated items used in festal processions and games, the spoils from the Persian Wars, and other resources of these kinds, all of which amounted to not less than five hundred talents. To this he added the money from other sanctuaries, which was not inconsiderable. And if they were ever compelled to such extremes, they could even use the gold that bedecked the goddess [Athena

Parthenos] herself. He informed them that there was a weight of fifty talents of pure gold on this statue, all of which could be removed. But he also pointed out that this gold could only be used for the exigency of their own self-preservation (*soteria*) and would afterwards have to be restored in the same or greater quantity.[16]

Fortunately, quite extensive records survive of the "resources" stored in the various rooms of the Parthenon and the Erechtheion during the period 434–295 BC.[17] To give at least a preliminary sense of the extraordinary scale and diversity of the contents of these inventories, a very selective list might include the following: hammers, anvils, and dies for pressing coins; coined and uncoined pieces of gold and silver; gold and silver incense burners; silver lamps; gilt wooden baskets and boxes; numerous gold and silver *hudriai* (water containers), *phialai* (bowls for liquid offerings), *oinokhoai* (wine jugs), and other ritual vessels; silver goblets and drinking horns; sacrificial knives; numerous gold wreaths; precious rings, earrings, necklaces, neckbands, headbands, and other jewelry; numerous sealstones; chitons and other robes; golden wheat stalks; gold, silver, ivory, and gilt wooden figurines; golden statues of Nike; ceremonial Chian couches, Milesian couches, stools with ivory backs, and thrones; gilt and ivory flutes, lyres, and other musical instruments; numerous shields, helmets, spears, arrows, swords, daggers, and other armaments and weapons, including items plundered from Persians and other adversaries.[18]

Perhaps one really could put a monetary value on such *hiera*, albeit an extraordinarily high one, as Pericles' speech suggests. But it also seems safe to say that they were collectively worth more to the Athenians and their patron goddess than the mere worth of the metals and other materials from which they might once have been made. When one considers the cumulative sum of the ritual meanings, the historical resonances, the artisanal skills and efforts, and, in some cases, the military sacrifices that made these myriad items the "treasures" they were at the time, their worth defied any modern-style economistic value standard.

Much the same might be said of the vast, ongoing traffic in sacrificial offerings that the Athenians shared with their gods, above all the animal offerings which were shared at communal banquets. But again, the purely material resources invested in such offerings can hardly be understated.[19] One fourth-century source claims that Demos spent more on "communal banquets and meat distributions" (*koinas hestiaseis kai kreanomias*) than on the everyday "management" (*dioikesin*) of the *polis*.[20] More generally, another notes how the Athenians were unusually prone to use poliadic resources to fund such events:

As for sacrifices, offerings, festivals, and consecrated lands, Demos recognizes that it is not possible for each poor Athenian by himself to conduct sacrifices, to feast, to give offerings, or to inhabit a beautiful and great *polis*, so it has found a way for this to come to pass. The *polis* thus makes numerous sacrificial offerings at communal expense. Demos is thus the one who feasts and shares the offerings.[21]

According to the most detailed study to date on the subject, a typical Athenian would have had the chance to take part in somewhere between forty and forty-five such ritual banquets per year, almost one per week. Festivals of the *polis* alone would have offered at least fifteen or sixteen such opportunities. Similar feasts staged in the demes offered perhaps an additional twenty or more. And one might have had access to perhaps five or ten more through one's tribe, phratry, *genos*, and the other such associations to which one belonged.[22]

While elites also staged sacrificial banquets at their own expense in their own homes, it is likely that the occasions for most of the feasts shared by poorer Athenians were the rituals celebrated in demes and by the *polis* as a whole.[23] To judge from surviving calendars, each deme would typically have held somewhere in the range of twenty to twenty-five such events each year. To give us some idea of the number of larger animal victims involved, we might cite the total annual figures pooled from three representative deme calendars:

	Oxen	Sheep/Goats	Swine
Erchia	o	39	o
Thorikos	3	40	o
Marathon	6	27.5	1
Total	*9*	*106.5*	*1*[24]

Of course, these figures represent sacrifices at just three of the 139 known demes. Since estimates suggest that members of these three demes combined totalled some 4.2% of the entire Demos, it can then be inferred that something like 214 oxen and 2,531 sheep and goats would have been sacrificed and consumed annually at deme banquets across the *polis* as a whole.[25]

The other major occasions for communal banquets that were attended by "ordinary" Athenians were larger *polis* festivals. Demos sponsored a range of events at which sacrificial meat was distributed. Some of these, like the "entry rites" performed each year when councilmen first took office, the offerings by the *prutaneis* ("presiders") at the start of each assembly session, and sacrifices made at meetings of the ten generals and other boards of *polis* officials, would have been relatively small in scale, with the meat shared only among the officials in question.[26] Other such events, like those included among the "ancestral sacrifices" (*patrioi thusiai*) which were listed in the recodified "calendar of Nikomakhos" around the end of the fifth century, also appear to have been comparatively modest affairs. Typically, no more than a single sheep was sacrificed, with meat distributed only among those who were responsible for actually conducting the rites.[27] But much larger meat distributions took place after the so-called *epithetoi heortai*, the festivals that were thought to have been "added" some time after the ancestral sacrifices were first established. This category included the Panathenaia, the City Dionysia, the Eleusinia, and other major festivals of the *polis*, which typically featured offerings of large numbers of oxen.

A general impression of the numbers of victims slaughtered and consumed on these occasions can be extrapolated from the "Dermatikon Accounts," partially preserved records of the revenues from oxhides sold by Demos after these festivals in the period 334/3–331/0.[28] As Vincent Rosivach has shown, if one assumes a constant 10 drachma value for an oxhide, the "highest likely price," one can then infer that the minimum number of victims sacrificed at each of the *epithetoi heortai* (where figures are preserved for one of the years) were roughly as follows:[29]

Sacrifice to Eirene (Peace)	71 oxen
Lesser Panathenaia	Missing
Eleusinia	Missing
Sacrifice to Demokratia	41 oxen
Asklepieia (Epidauria)	100 oxen
Theseia	118 oxen
Dionysia in Piraeus	31 oxen
Dionysia at the Lenaion (Lenaia)	34 oxen
Sacrifice to Agathe Tyche (Good Fortune)	10 oxen
Asklepieia	24 oxen
City Dionysia	81 oxen
Olympieia	63 oxen
Sacrifice to Hermes Hegemonios	Missing
Bendideia	46 oxen
Sacrifice to Zeus Soter	105 oxen
Total	*724 oxen*

Adding to this total the one hundred oxen which are known to have been sacrificed annually at the Lesser Panathenaia, along with the eight or nine that were awarded as prizes at the festival, Rosivach then arrives at a total minimum figure of 832 oxen that were ritually slaughtered and consumed per annum at *epithetoi heortai.* Moreover, as he notes, this already substantial figure would have risen even higher sometime after 338/7 BC, when the *polis* allotted 41 mnai for further victims for the Lesser Panathenaia. Since that sum was sufficient to pay for at least 41 more animals, this would give us a new total of 873 oxen. Furthermore, this figure still would not include the unknown numbers offered at the Eleusinia or the sacrifice to Hermes Hegemonios. Nor would it take into account the additional numbers of victims sacrificed at the six quadrennial or "penteteric" festivals, events like the Great Panathenaia, the Delian *theoria,* and the Brauronia, all of which featured communal dining.[30] Nor for that matter would it include the five hundred goats killed at the annual sacrifice for Artemis Agrotera, the one *epithetos heorte* that did not include offerings of oxen, and was thus excluded from the Dermatikon Accounts.[31]

If we then add together all the above figures for animals sacrificed by demes and by the *polis* as a whole, we arrive at the tentative conclusion that the Athenians each year sacrificed at least 3,031 (2,531 + 500) sheep and goats, and 1,046 oxen (214 + 832), with the number of the latter rising to something like 1,087 after 338/7 BC.[32] So where did all these victims come from? Who reared and sold them? And how much did they cost?

There is no evidence that the demes or the *polis* itself maintained stocks of sacrificial animals. Given the very large numbers of sheep, goats, and oxen that were ritually slaughtered and consumed annually, it seems likely that the victims were acquired from the relatively small number of wealthy Athenian *oikoi* that could afford to maintain large flocks and herds.[33] The animals in question were not raised expressly for their meat, sacrificial or otherwise. The typically "mature" victims purchased from these sources would have been raised originally for dairy goods, wool, and (in the case of oxen) for service as draft animals, so would only have been sold for sacrifice towards the end of their working lives.[34] Sales of victims to the demes and *polis* would thus have been only a secondary source of income for the sellers. That said, this income may still have been quite substantial. Again using Rosivach's figures, the possible ranges of total annual expenditures on animal sacrifices by demes and *polis* can be tabulated as follows:[35]

	Number	Cost per head	Total cost
Sheep/goats	3,031	10-17 dr.	30,310-51,527 dr.
Oxen	1,046	50-100 dr.	52,300-104,600 dr.
Total			*82,610-156,127 dr.*
			= 14-26 talents

To put it another way, if, for the sake of argument, we assume an average annual total outlay of some twenty talents on victims, the demes and *polis* would then have spent roughly 300 talents on animal sacrifices during the fifteen years it took to complete the work on the Parthenon, or well over three-fifths of the cost of the building itself. And given that our figures necessarily exclude the costs for the annual Eleusinia, the sacrifice to Hermes Hegemonios, and the major penteteric festivals, we can safely assume that the total annual expenditures on ritual victims in Attica would have been considerably higher still.[36]

So where finally did the funds for all these extravagant outlays on temples, treasures, and sacrifices for the gods come from?

Obviously, the contents of the treasury rooms on the acropolis came from a wide variety of sources. They included battlefield spoils, dedications by individual families, and, one suspects, a good number of time-honored ceremonial items whose precise origins had long been forgotten by the time they came to be stored in the Parthenon.[37] But probably the majority of at least the more recent treasures were paid for ultimately out of the general funds of Demos, whose regular revenues derived largely from the lease of *polis* holdings, such as the silver mines, from various fiscal impositions, like the *metoikion* paid by metics, import and export duties, and occasional *eisphora* levies on the wealthy, and from court fines and confiscations.[38]

Likewise, these same common funds would have been used to fund major temples. As Lisa Kallet has persuasively argued, even the Parthenon, that proverbial trophy of empire, was in all likelihood funded for the most part from the Treasury of Athena, not directly from imperial "contributions."[39] And it is no less likely that most of the money spent by Demos on *polis* sacrifices came from the same source. While it appears that the modest numbers of victims required by the *patrioi thusiai* were paid for from leases of the lands (*temene*) belonging to the gods in question, the far larger numbers of victims required by the *epithetoi heortai* were again funded by regular *polis* revenues, perhaps supported by the resources of special reserves like the Theoric Fund.[40] As for the costs of deme sacrifices, they were probably covered either by wealthy demesmen as a benefaction or by general deme revenues, which derived from sources like occasional taxes, land leases, fines, and interest on loans made to members.[41]

All of this in turn helps us to recognize the quite extraordinary extent to which the Athenian social body depended on more prosperous households to fund and sustain its vital, life-sustaining exchanges with its gods.[42] The vast majority of all of the ritual outlays discussed above relied upon outlays of resources, direct and indirect, by affluent *oikoi*, both Athenian and non-Athenian. Wealthy families controlled the large flocks and herds that probably supplied most if not all of the sacrificial victims required by the demes and the *polis*. Wealthy families must have paid for most if not all of the personal dedications which were numbered among the acropolis "treasures." Wealthy families were of course solely responsible for the liturgical outlays which were also essential to the performance of at least ten major poliadic festivals and an unknowable number of others in the demes.[43] And more generally, as the primary contributors of taxes, lessees of communal lands, mines, and *temene*, the most significant payers of import and export duties, and as the principal payers of larger court fines and victims of confiscations, wealthy families were directly or indirectly responsible for most of the regular revenues which the *polis* itself and its demes used to pay for so many of the gifts they bestowed on their gods. In other words, the reproduction of the most basic and essential conditions of Athenian existence depended very largely on the resources of a small, affluent minority.

It may of course be possible to weave all the kinds of data discussed above into some kind of story about an "Athenian economy." But to be theoretically meaningful as such and self-consistent, this would necessarily be a secular story, a story exclusively about resources that are deemed to possess a narrowly economistic value, a story about the average and aggregate actions of human individuals. Which is to say, it would be a story whose historical value would be questionable, to say the very least. For this story would necessarily exclude the most important parties in all of the exchanges just described, namely the gods themselves. And it would necessarily exclude all gifts which the gods bestowed upon the Athenians in return, namely all the essential conditions of their existence. More generally, by reducing ritual actions effectively to economic transactions, it would necessarily exclude as "non-economic" all of the affective, experiential, and otherwise less readily quantifiable dimensions of Athenian exchanges with divinities, dimensions which gave those exchanges much of their meaning and significance at the time.

For as Theodora Suk Fong Jim has recently emphasized, it would be a mistake to see Greek gift-giving to divinity simply as the expression of some kind of quasi-contractual, *do ut des* ("I give so that you may give") relationship, one founded purely on the mutual expectation of some future material return. Greeks shared a deeper, more complex, more constant and taken-for-granted sense of *kharis* ("reciprocity") with their gods, routinely giving what amounted to thank offerings for past services rendered, such as "first fruits" (*aparkhai*), "tithes" (*dekatai*), and dedications made before meals, after battles, and so forth.[44] Given the inevitable asymmetry of human exchanges with divinity, this sense of reciprocity surely did involve something like feelings of personal "indebtedness" and "dependence" on gods, as Jim has observed. And it also no less surely involved a certain kind of sociality, a sense of communion that could include "elements of mutual goodwill, pleasure, and delight."[45]

For the Athenians, the gods of Attica were not some group of faceless, super-human hired contractors. They were something closer to benevolent governors or caring parents, beings who perpetually monitored and managed the shared local environment, taking a personal interest in the life and well-being of their chosen people.

In the past, for example, as we see in tragedies like Aeschylus' *Eumenides*, gods had intervened decisively to resolve human disputes and establish procedures and institutions which would help to secure peace and harmony in the future. They could also facilitate military victories, as they did in 403, when they engineered the battlefield conditions which allowed the exiles to defeat the forces of the Thirty, thereby helping to restore the integrity and rule of Demos.[46] As we see in the ephebes' oath, they routinely came to witness the incorporation of Athenian youths into the social body. And as we see in the east section of the Parthenon frieze, they convened to watch the lavish ritual spectacles of the *polis*.

Indeed, as Pindar's well-known dithyramb for the Athenians vividly illustrates, gods could also be expected to play a more active role in the celebration of their fellow divinities. In this instance, they are invited in early spring to join in a choral dance for Dionysus, an event which appears to have been staged in the area of the old "Cecropian agora," the neighborhood that contained the eternal fire of Hestia, the sanctuary of Aglauros, and other such time-hallowed sites.

[Come] here to the chorus and send glorious grace upon it, Olympian gods, you who approach the much-trodden, fragrant-with-incense navel-stone of the city in holy Athens and the all-decorated, famous agora. Receive a share of crowns bound with violets and songs culled in the spring, and look [with favor] upon me as I go from Zeus with the radiance of songs secondly to the ivy-knowing god, whom we mortals call Roarer (Bromios), whom we call Loud-shouter (Eriboas), singing and dancing in celebration of the offspring of the highest fathers and Kadmeian women. And clear [signs] do not escape my notice, as if I were a seer, when, with the chamber of the red-robed Horai opened, nectareous plants lead on the spring so that it is [even more] fragrant. Then, the lovely locks of violets are cast upon the ambrosial earth, and roses are mixed with hair, and voices of songs resound with the accompaniment of pipes, and choruses approach Semele with her circular headband.[47]

As two recent observers have said of this text, "everything here is about fusion and integration of all the different elements—gods, mortals, and nature—united in ecstatic Dionysiac worship."[48] And as we have observed throughout the latter part of this study, such cosmic integration was not just a characteristic of this particular past world, but the ultimate condition of its very existence. The secular, scientific language of "economy," I submit, cannot begin to capture the nature or significance of the resource exchanges that were required to sustain such a world. The more inclusive, less irredeemably modernist language of "ecology," a language that might allow us to better represent the many valences of human exchange relations with the gods and the land of Attica, would seem to be a good deal more suitable.

This alternative account of human-divine exchanges offers just one illustration of the ways in which a more inclusive, recursive approach can change how we think about the larger circuitry of life-sustaining resources in Attica. In the remainder of the chapter, we can look more summarily at three further illustrations.

Resourceful Women

First, a recursive analysis of resource flows in Attica would help us to appreciate even more the vital significance of women to the life of the social body. In a sense, females were themselves resources. And probably the most vital, most ecologically essential exchanges between humans in Attica were not sales of any produce or commodities but marriages.

Athenian marriages involved complex resource transactions between two *oikoi*. When a bride moved from her natal to her marital *oikos*, her family received from the groom's family nothing less than a commitment to maintain and protect her for the remainder of her life. In return, the groom's family received a range of different resources. Especially in the case of marriages between more properous Athenians, these resources began with a dowry (*proïx*), which could come in any number of forms, from money and jewels to furniture and sometimes land. Though reclaimable in the event of divorce or widowhood, this dowry served as the wife's pre-mortem inheritance, thereby materially uniting the resources of the two *oikoi* and underwriting her standing as a full partner in the new marriage.[49] At the same time, of course, the groom's family also received the vital benefits of the wife's natural spousal and maternal dispositions, the kind of divinely bestowed female capacities and aptitudes for managing a household that were discussed in Chapter 12. And last but not least, the groom's family received the no less vital reproductive resources of her body. The precise nature and implications of these corporeal resources deserve a little further attention.

If one were to judge by a small handful of well-known texts, one might be led to believe that Athenians and other Greeks minimized the contribution of women to the physical formation of the embryo. In these sources, the female womb is assumed to be no more than a "field" to be "ploughed" or a kind of incubator for the seed of a new person that has been implanted by the male.[50] But

as Lesley Dean-Jones has emphasized, a larger number of texts, including pertinent examples in the Hippocratic corpus, tell a rather different story, whereby
the embryo is a product of a fusion between male and female seeds, even if the
exact nature of the latter is not always understood in the same way by different
authors.[51] And in all likelihood, she contends, this latter theory was "predomi-
nant" in archaic and classical Greece, representing the everyday, common sense
account of the time.[52]

Thus the reproductive resources of the wife's body played a critical role in the
very material constitution of the children that would perpetuate her husband's
patrilineal *oikos*. Each child in effect replicated part of the essential self of each of
the two parents. As Jérôme Wilgaux has put it:

> For a father and a mother, the newborn child is like a piece of their own
> flesh which comes apart to form a distinct and yet similar person. . . .
> Through the procreative process, the embryo, fetus, and child becomes
> the material and formal continuity of his genitors.[53]

Hence, one could expect to see the physical characteristics of both parents manifest in their offspring.[54] Indeed, children were even thought to replicate the behavioral traits of both their mothers and their fathers.[55] Likewise, it was taken
for granted they could inherit hereditary forms of pollution from either parent,
as is attested by the famous case of Pericles, who was deemed by the Spartans to
bear the ancestral curse of the Alcmeonid *genos*, to which he was related on his
mother's side.[56] But it is perhaps in attitudes towards certain illicit sexual unions
(*anosioi gamoi*) that we see the strongest evidence for the normativity in classical
Athens of the "two seed" theory, which rendered the female contribution to conception decisive. As Dean-Jones points out:

> [D]espite the patrilineal structure of Greek society in which children
> were part of the family of their father and not their mother, Athenian law
> allowed the marriage of half-siblings from the same father but not those
> from the same mother: originating from the same womb made such a rela-
> tionship incestuous where springing from the same seed did not.

The assumption in classical Athens that there prevailed what Wilgaux has
termed a "consubstantiality" between children and both of their parents duly
helps to explain why the Athenians came to insist on endogamy within the
polis. One could only be a true Athenian if one was oneself fashioned from the
living tissues of a male and a female Athenian. More to the point, Athenian
thinking about conception and procreation only deepens our sense of the critical

contributions made by female resources to the overall life and well-being of the *polis*. It was not just the case that women were self-evidently essential to the reproduction and management of every individual Athenian *oikos*. As bearers of the wombs where the material fusion of the essences of different *oikoi* took place, women collectively provided the sites where the very unity and continuity of Demos as a living social organism were ultimately forged.

Elite Contributions to the Life of the Social Body

As the opening section of this chapter has already intimated, viewing Attic resource flows on their own local ecological terms also allows us to see the truly remarkable extent to which the reproduction of Athenian social being depended upon the multifarious contributions of the wealthy minority. Nor should this surprise us. If the *polis* was a social body of households, the great majority of which possessed little more than the resources necessary to reproduce themselves, it was natural and inevitable that the task of sustaining the life of the totality, the life of the corporate Demos itself, fell heavily upon those few households who maintained a significant surplus. Recalling from the previous chapter the three areas of ecological concern over which Demos exercised a particular competence, we might now consider in turn the contributions made by elite resources to the management of Athenian *hiera*, relations with peoples elsewhere, and the shared *hosia* of the *polis*.

Consider first the contributions made by elite households to the ritual activities which secured the "health and safety" of the *polis*. As we saw a little earlier, these contributions went far beyond the familiar "liturgies," where individual families directly funded, say, choruses at drama festivals or teams at athletic spectacles. Even the major gifts of the *polis* to divinity, like temples, treasures, and animal sacrifices, gifts which were funded from what we conventionally call the "public" reserves of Demos and the demes, were in the end largely drawn from the "private" resources of the wealthy few.

Moreover, the Athenians benefited continually and directly from the special relationships which particular superhuman agencies enjoyed with individual elite human *oikoi* and *gene*, like the Boutadai, who had always supplied all the priestesses for Athena Polias and the priests for Poseidon-Erechtheus; like the Eumolpidai and the Kerykes, who had been custodians of Demeter's mystery cult at Eleusis since time immemorial; like the Phytalidai, who had once cleansed Theseus of his blood pollution and provided all the priests for his cult thereafter; and like the family of the tragedian Sophocles, who actually welcomed the god Asclepius to their own house when he first arrived in Attica.[57] Likewise, the Athenians benefited directly from those distinguished *gene* who regularly acted as official *theoroi* (ritual representatives) of the *polis* at major sanctuaries elsewhere, like those of Apollo at Delphi and Delos.[58] More generally, they benefited from all the god-pleasing festive spectacles that were made possible by members of more prosperous *oikoi*, who alone possessed the education, training, and skills to serve as playwrights, chorus members, athletes, horsemen, chariot-dismounters, armed dancers, and so forth. And they particularly benefited from those elite families whose female members traditionally assumed key roles in some of the most eco-logically critical rituals of the *polis*, whether helping to weave Athena's *peplos* as *arrhephoroi*, performing as she-bears for Artemis in the *arkteia*, or leading the massed ranks of the great procession at the Panathenaia.

Much the same could be said of the various resource contributions made by elite households to the management of relations with peoples elsewhere. The most urgent of these were of course the contributions made to military opera-tions and the defense of the *polis*. In a world where land combatants were re-quired to provide their own arms and equipment, it was simply expected that the more prosperous Athenians, perhaps the wealthiest 25–30%, would supply the necessary heavy infantrymen for the phalanx. It was similarly taken for granted that a far smaller, conspicuously wealthy minority would furnish all the necessary cavalrymen (*hippeis*) and mounted archers (*hippotoxotai*), whether they had to loan the considerable cost of a horse from the *polis* or owned their own mounts outright.[59] And it was taken for granted that only the most affluent would be responsible for paying all occasional warfare-related "contributions" (*eisphorai* or *proeisphorai*) and of course for covering the enormous liturgical costs of maintaining and manning the two hundred or more triremes of the all-important navy.[60]

Nor were elite contributions to the defense of the *polis* limited to outlays of material resources. When Demos assembled to consider military strategies and logisitics, outcomes were necessarily framed and determined by members of the tiny "liturgical class" minority, who alone possessed the requisite knowledge and education to formulate viable proposals. For much the same reasons, the delegations that were sent by Demos to other human communities to negotiate alliances, treaties, and other agreements were like-wise invariably dominated by this same tiny minority. Of course, elites also largely monopolized all positions of battlefield leadership, like the ten annual generalships, the hipparchies, the phylarchies, and the taxiarchies. And even during campaigns, the war efforts of Demos could still benefit further from the material resources of its leaders, since generals were sometimes expected to provide supplemental funds from their own personal holdings or from per-sonal loans, especially in the fourth century when military operations were sometimes seriously under-financed.[61]

If elite households were thus ultimately responsible for providing most of the resources, fiscal and otherwise, that Demos needed to fund, plan, and manage *ta hiera* of the *polis* and relations with other peoples, both friends and enemies, these same households also made critical contributions to the maintenance of *ta hosia* in Athens.

At least in the early classical era, before about 450 BC, it was apparently not uncommon for individual *oikoi* to provide communal services and facilities of their own accord. To take only the best known example, in the two decades after the Persian Wars, the proverbially wealthy Cimon was credited with numerous acts of personal *megaloprepeia* ("munificence"), from helping to fund the Long Walls and beautification projects in the agora and Academy to providing daily sustenance to needy neighbors in his deme of Lakiadai.[62] And even thereafter, when general *polis* funds were routinely used to cover the costs of constructing and maintaining all major communal buildings, roads, and other shared utilities, it is important to remember again that the bulk of these funds derived ultimately from the resources of more affluent families, those who leased silver mines, paid duties on imports and exports, paid significant court fines, and so forth.[63]

Furthermore, throughout the classical era, prosperous Athenians were continually responsible for securing and managing supplies to the *polis* of grain, most if not all of which had to be imported. As Alfonso Moreno has quite conclusively demonstrated, "an elite of Athenian politicians" effectively "controlled" this grain supply through their "powers of leadership, organization, distribution and persuasion."[64] In the fifth century, the social body's grain needs were typically met by the surpluses produced by large, elite-controlled "clerouchies," lands formerly belonging to other *poleis* which had been forcefully appropriated and cultivated by Athenians. In the fourth century, following the demise of most of these clerouchies, grain was imported primarily by arrangement with the Spartocid dynasty, lords of the Bosporan kingdom, which controlled extensive lands in the Black Sea region. Either way, Moreno shows, the provision of this most basic and essential of all foodstuffs was at all times dependent upon the resources of a relatively small number of elites, who served as principal "advisors" to Demos, as diplomatic negotiators, as managers of large clerouchies, as ship-owners, as grain-importers, and as grain-dealers in Athens itself. And their various cumulative efforts must be judged a significant success, given the remarkably low number of grain shortages suffered by the *polis* during the period covered by this book.[65]

Last but not least, as previous sections have already indicated, one can hardly understate the degree to which all forms of communal decision-making process in the *polis* were dependent upon resources of the wealthiest, most educated Athenians. Again, it was clearly taken for granted that such men would assume responsibility for prosecuting all cases where the well-being of the *polis* as a whole at stake. And again, we do not know of a single significant assembly "speaker" or "advisor" who did not demonstrably belong to the "liturgical class." While the "power" enjoyed by even the most influential of these figures may in the end have been only *de facto* in nature, few would question Finley's conclusion that they nonetheless together constituted an essential "structural element" in the deliberative process, bringing to debates a degree of leadership, rhetorical skill, knowledge, and vision that would otherwise have been missing.[66] The very nature of decision-making mechanisms in Athens thus presupposed the availability of such elite resources.

None of this is to belittle the contributions made to *polis* life in general by non-elite Athenians, the great majority of whom were surely able to sustain their own *oikoi*, some by providing useful goods and services of various kinds to their fellow Athenians. It is rather to highlight the extraordinary degree to which the life of Demos itself, the ontologically distinct life of them all as a corporate self and subject, was dependent upon the manifold surplus resources that only elite households could provide. Yes, non-elites also made various contributions to this corporate life, as ritual actors, troops, rowers, builders, assemblymen, judges, and *polis* functionaries, even if they often had to be paid small stipends in return. But without all of the elite contributions, both material and immaterial, the classical Athenian *politeia* as we know it would have been unfeasible if not unthinkable.

Even then, one might still contend that at least some of these resource flows would be consistent with the consensus vision of a "democratic Athens." Despite the manifest dependency of the *politeia* upon a highly unequal distribution of essential resources, despite all the self-evident entanglements between the health of the social body and the vitality of its wealthier households, one might still argue that financial outlays on liturgies and the like supported a general commitment to egalitarianism in the *polis*. For in effect, such expenditures might have amounted to a kind of "economic redistribution" from rich to poor, thereby helping to assuage possible class-like "tensions" between the two.[67] But even if there is a measure of truth in this claim, its force should be qualified by at least four further considerations.

First, the purpose of liturgies and other such outlays was not to "redistribute" wealth from rich to poor or otherwise temper inequalities between families or individuals. It was rather to secure the life and well-being of the social body as a whole, a body which included all Athenians, rich and poor alike. Second, this redistribution idea again presupposes a narrowly economistic view of the resource transfers in question. In so doing, it minimizes or overlooks the more intangible but no less essential resources that accrued to elites as a result of all these "gifts" to Demos, resources like *kharis, time* ("honor"), *arete* ("excellence"), *eusebeia* ("piety"), and *dikaiosune* ("sense of fairness"), along with the communal influence in decision-making that one might enjoy as result of possessing such virtues.[68] Just as there was a natural and inevitable asymmetry in the exchange relations between Demos and the gods, a relational asymmetry no less naturally and inevitably prevailed between rich and poor Athenians, such that the latter cannot but have felt an almost constant, taken-for-granted sense of dependence, even indebtedness towards the principal leaders and sponsors of the *polis*.

Third, elite contributions to the life of the social body went far beyond liturgies and other explicitly financial outlays. As we have seen in this chapter, they assumed all manner of forms, some readily observable, some less so. And they affected almost every fiber of Athenian social being, from the production of sacrificial animals and the management of cults to assembly deliberations, the resolution of legal disputes, the exigencies of defence and diplomacy, and the provision of grain for the *polis*. Fourth, and most fundamentally, to claim that elite expenditures somehow served the cause of an egalitarian *politeia* would again be to ignore the distinctive metaphysical environment in classical Athens. In a world where social being, not individual being, was primordial, where gods had not made humans equal or interchangeable, where the polity was a pregiven communion of unequal but interdependent households, it was natural and inevitable that those who possessed significant surplus resources were obliged to share them with those who did not, thereby benefiting the whole. They gave most to the life of the social body because they had benefited most from this life and thus had the most to lose from its extinction. And for all the *hiera* and *hosia* of the *polis* that their surplus contributions made possible, it was not as if they received nothing in return.

Finally, drawing attention in this way to the extraordinary dependence of Athenian social being on elite resources in turn prompts us to consider afresh one further major resource flow in the *polis*, one that furnished most of the surplus revenues which made elites elite in the first place. I refer to the exploitation of slaves.

Slavery and Athenian Social Being

As we also observed earlier, there is much we do not know and can never fully know about the experience of slavery in ancient Athens. We have only the vaguest idea about the early history of the use of imported chattel slaves in Athens. And even a general knowledge of the total numbers of slaves in fifth- and fourth-century Athens continues to elude us. But again few would doubt that these numbers amounted to at least around 80,000 by the time of the Peloponnesian War, meaning that slaves made up somewhere between a quarter and a third of all the human inhabitants of Attica, proportions comparable to those attested in the ante-bellum South of the United States.[69] Few too would doubt that slaves owned in common by Demos made essential contributions to the everyday management of the *polis*.[70] And no one doubts for a moment that households made extensive use of slave labor in all manner of activities, both productive and non-productive, but especially in mining, in larger-scale commodity production, and in agriculture on larger landed estates. So few presumably would deny that Athenian experience supports Moses Finley's general observation that slavery was a "basic element in Greek civilisation."[71]

More important for our immediate purposes, few would doubt that the Athenians who made the most extensive use of slave labor were the wealthy. Given that the going rate for a single highly productive or skilled male slave was probably well over 200 drachmas, we can safely assume that ownership was quite heavily concentrated among the more prosperous *oikoi*.[72] Moreover, in a world that knew nothing of commercial corporations, a stock market, or the technologies of automated mass production, a world where there was little possibility of employing fellow members of one's own social body on anything other than a casual or occasional basis, the only way one's household could accumulate significant surplus resources in activities like agriculture, mining, and commodity production was through the exploitation of workers imported from elsewhere. As Geoffrey de Ste Croix once put it, "there was simply no way in which the propertied classes of the Greek world could obtain a substantial surplus directly except through unfree labour."[73] And given that the most lucrative non-productive profit-making pursuits, such as banking, insurance, and trade, ultimately presupposed the existence of prior surpluses from more directly productive activities, it is very hard to see how family fortunes of any size were even conceivable in classical Athens without the wide circulation of profits from slavery.

How many unfree laborers, then, were actually owned by the more affluent surplus-producing families? Plato apparently took it for granted that it was the mark of a distinctly "wealthy" (*plousios*) household head to own at least fifty slaves.[74] And this figure seems to be generally supported by other evidence.

Consider, for example, the case of the *oikos* of the celebrated *rhetor* Demosthenes. Though his ancestral household was far from the wealthiest in Athens, the orator's father was still apparently able to bequeath him a fortune of almost fourteen talents (84,000 drachmas), a sum which would place its holder firmly in the liturgical class. And apparently the principal sources of this wealth were workshops for knife-making and couch-making, which between them employed 52 or 53 slaves.[75] Yet these operations were relatively modest compared to the shield-making workshop that the speechwriter Lysias and his brothers inherited from their father Cephalus, which employed as many as 120 slaves before the family was persecuted by the Thirty. Though these men were metics, originally from Syracuse, their considerable wealth had previously allowed them to move easily in elite Athenian circles and had rendered them fully liable to pay special levies (*eisphorai*) and perform multiple liturgies as chorus-sponsors at festivals.[76]

The most detailed picture we have of the operations of a large agricultural estate comes from the pages of Xenophon's *Oeconomicus* ("Household Manager"), where the household in question is again that of the Ischomachus mentioned in Chapter 12. While we are not told exactly how many slaves Ischomachus possessed at the time, the numbers were sufficiently large that he himself had to train foremen (*epitropoi*), who were themselves most likely slaves, to supervise them all in the necessary tasks of "planting, clearing, sowing, and harvesting" on the estate.[77] And apparently this estate was extremely lucrative, since Ischomachus performed frequent liturgies and was said to have accumulated a huge fortune of some seventy talents (420,000 drachmas).[78]

More lucrative still, it seems, was the business of leasing one's slaves out for work in the mines. As we learn from another work of Xenophon, an otherwise unknown Philemonides possessed no fewer than two hundred slaves which he leased out for mining, earning him 30 drachmas per day. Then again, twice this amount was reportedly earned daily by the six hundred slaves of the more illustrious Hipponikos, proverbially the richest man in Greece, who was a member of the Kerykes *genos*, a descendant of a line of Olympic equestrian victors, and an heir to the colossal 200 talent (1,200,000 drachmas) fortune of his father, the proverbially "pit-wealthy" (*lakkoploutos*) Kallias. And more than 160 drachmas per day were apparently earned from this same source by Nicias, the eponymous architect of the short-lived peace with Sparta in 421 and commander of the ill-fated Sicilian campaign in 415–413. Nicias, it seems, possessed an astonishing one thousand slaves and a net wealth of 100 talents (600,000 drachmas), roughly enough to sustain an entire deme of one hundred smaller families at around subsistence level for fully twenty years.[79]

Given such evidence, it is again certainly possible to tell stories about the larger impact of slavery on classical Athens using conventional economistic forms of analysis, whether liberal or orthodox Marxist. One could readily draw attention to the significant role played by slave labor in the Athenian "economy" as a whole, especially in "sectors" like mining, commodity production, and commercial agriculture. At the same time, like the self-identified Marxist Ste Croix, one could argue that the existence of a socially dominant "propertied class" in classical Athens was very largely dependent upon the exploitation of unfree labor. And one could even claim, as some have done, that slavery made possible the very practice of "democracy," in that it gave large numbers of Athenians the leisure to serve as assemblymen, councilmen, judges, and so on.[80]

But such stories would again all take for granted a disenchanted, individualist, proto-capitalist model of social being. If we instead embrace the recursive approach proposed in this study, using a local Athenian model of social being, one that thus encourages us to focus on wider circulations of resources within a social body of interdependent households, not merely on aggregate transactions between individuals, we can begin to tell a very different and perhaps more revealing story. For we can begin to see that the very life of this social body as a whole, what the Athenians called their *politeia*, their *demokratia*, was all but unsustainable without the profits from slave labor. Again, given that the members of Demos did not work for one another as a rule, practically the only way for their households to generate the surplus resources necessary to reproduce this Demos, to sustain its existence as a discrete corporate self, was through the exploitation of workers imported from elsewhere.

Slavery then was not just a "basic element" in Athenian "civilization." Without too much exaggeration, one could even say that the entire ecology of the classical Athenian *polis*, and thus the very life of the social body, depended on slavery. For what Marx would have called the "dead labor" of slaves, the surplus value extracted from their toils, was inscribed and impressed in all the mechanisms and practices of Athenian *demokratia*. It was immanent in the very stones of acropolis temples, in the meat and bones of sacrificial victims, in the shining metals of votive *hiera*, in the timbers of triremes, in the bronze of hoplite panoplies, in the calculations of military strategies, in commands given on battlefields, in the contents of assembly resolutions and courtroom verdicts, and in the very bread that nourished all in Attica.

Before closing, one might ask: Were there any other modes of surplus accumulation available at the time that could have generated the quantities of resources necessary to sustain the life of Demos? To a point there was one, and it was one that the Athenians did also pursue, albeit just for around fifty years in the latter half of the fifth century. I refer of course to what is commonly called the "Athenian empire," whereby very significant surplus resources were regularly extracted from the social bodies of some 170 other *poleis*. Doubtless, the revenues from this imperial project helped to alleviate some of the more extreme burdens of sustaining social being that had hitherto fallen upon elite *oikoi*. Doubtless too, the use of imperial forces to secure Athenian control of the Hellespont region and most of the Aegean encouraged new flows of resources into the *polis*, especially through those prosperous families who were now able to maintain large clerouchies and exploit other previously unavailable commercial opportunities.

Nonetheless, this fifty-year enterprise did not permanently or irrevocably alter the essential nature of the ecology that prevailed in Attica itself.[81] After all, at least in principle, imperial revenues were reserved exclusively for imperial purposes, above all the purpose of protecting the security and interests of the empire itself and its many subjects. Accordingly, until late in the Peloponnesian War, these revenues were kept separate from the general funds of the *polis* and administered by their own managers, the board of Hellenotamiai ("treasurers of the Greeks"), even if the Athenians sometimes drew on this reserve to repay "loans" from other treasuries. Meanwhile, those other treasuries, principally that of Athena and, latterly, that of the Other Gods, still harbored most of the revenues that were required for the everyday management of the *polis*, both for covering everyday expenses and even for funding special projects like the Parthenon, as we saw earlier.[82] These latter revenues still came predominantly from the usual sources, like rents, fines, and harbor duties, that were paid mostly by the wealthy. And even at the height of the imperial era, elites were still expected to perform liturgies and pay levies like *eisphorai* when the occasion arose.

In sum, the basic fabric of the ecology presented in this chapter, an ecology that was critically dependent upon the surplus resources produced by slave-holding elites, continued more or less unchanged down to the end of the classical era, and indeed for a long while thereafter. Without the wealthy and their slaves, it seems, there was no *demokratia* as the Athenians knew it.

16

Being in a Different World

PART II OF this study presented ethical and the philosophical cases for an alternative, non-dualist form of historicism, one that can support an ontological turn in the discipline. In the preceding chapters of Part III, the book has sought to demonstrate the practical, analytical benefits of this paradigm shift. In the end, these chapters have been trying to answer a single question: What would the world have to be like for the Athenian *politeia* to make sense as a mode of life? The net outcome is an emerging vision of a very different Athens, a distinctly non-modern Athens, where the prevailing world-making common sense sustained a way of being human that was profoundly different from our own.

A Non-Modern Humanity

Unlike the "democratic Athens" found in the modern literature, this non-modern Athens thoroughly confounds the ontological template of our conventional historicism. It was a world that could not and did not divide its experience into human and non-human, sacred and secular, or public and private realms. It was a world wholly unacquainted with distinct political, social, or economic fields. It was a world innocent of religion per se, as a discrete, bounded thing-in-itself. And it knew nothing of democracy, citizenship, or even a natural individual, at least as we understand those terms today. No equivalents of these modern analytical categories were either known or knowable at the time, because the uniquely modern metaphysical conditions of their possibility as phenomena were entirely missing from Greek antiquity.

So what kind of world do we see when we instead analyze Athens recursively, in its own distinctive metaphysical conjuncture? We see an Athens where social being not individual being was primordial, where "the Demos of the Athenians" was an ontologically primitive body of interdependent households, a kind of unitary human superorganism, not an ever-changing aggregate or "mass" of self-interested, mostly non-elite male "citizens." We see an Athens where women were not systematically subordinated or oppressed, but continually expected to perform vital, divinely ordained roles in securing the well-being and continuity of their households and their *polis*. We see an Athens where there was no free-standing apparatus of "government," where all members of the social body, male and female, were expected to manage their own lives and livelihoods from one day to the next. And we see an Athens where this social body only assumed visible, material form, as the unitary totality of Demos, to resolve matters that its myriad constituent parts, acting individually and collectively, could not already resolve for themselves.

At the same time, we see an Athens that was very far from egalitarian, an Athens where inequalities between members of the social body were fundamentally acceptable, since the primary mechanisms for reproducing that body were all crucially and self-evidently dependent upon the surplus resources produced by a tiny minority of super-rich households. We see an Athens where these vital surplus resources were routinely and necessarily extracted from the labor of slaves. And perhaps above all, we see an Athens where humans were not alone. We see an Athens where all of the practices of the *politeia* were ultimately just the human contributions to a more inclusive system of life, to a timeless cosmic ecology that forever bound Demos in a condition of symbiotic reciprocity with the land and the gods of Attica. Which is to say, we see an Athens where the preeminent beings in the *polis* were not actually the Athenians themselves.

It is of course for the reader to decide whether or not this inaugural exercise in ontological history yields an account of the Athenian *politeia* that is more historically meaningful than those found in the standard literature. Either way, this alternative, non-dualist account directly challenges the common assumption that one cannot write histories of non-modern peoples without using conventional social scientific master categories. And in so doing, it helps us to resolve a number of long-standing issues.

Most immediately, this alternative account of Athenian social being helps us to move beyond so many of the basic "category problems" which continue to consume so much scholarly energy. It allows us to see that the problems of determining, say, the precise nature of an Athenian "economy," the precise quiddity of an Athenian "religion," or the precise boundary between "public" and "private" spheres, are all in fact non-problems. Since all of the social objects in question presuppose a modern metaphysical conjuncture, none could possibly have been there in Athens in any sensible shape or form at the time.

In much the same way, the new account allows us to see that all of the alleged contradictions of the Athenian *politeia*, especially those between a secular, rational, egalitarian "democracy" and so many other ostensibly non-secular, ir-rational, unegalitarian, and "undemocratic" practices, are ultimately problems of our own making. For as we have now seen, the prevailing Athenian "model" of social being was entirely compatible with, say, a highly gendered division of soci-etal labor, with large wealth inequalities between the rich minority and the poor majority, with the extensive use of slave labor, and with lavish outlays on ritual activities. Indeed, such putative "problems" were all in their different ways essen-tial to the reproduction of the life of the *polis*. Far from contradicting *demokratia*, they were conditions of its very possibility.

Finally, by restoring the *polis* to its own distinctly non-modern metaphysical conjuncture and divesting it of the above "problems," the new account in effect liberates Athens from that proverbial modern metanarrative which inevitably infantilizes a celebrated ancient lifeworld as a perpetually immature anticipa-tion of our own. It actively resists our historicist predisposition to see Athens as a world full of absences, incompletenesses, and inadequacies, as a world con-founded by problems and contradictions that await their full resolution in a post-Enlightenment modernity. Instead, it invites and encourages us to see what Athens might look like when it is released from the burden of a destiny imposed by others, when it is just a fully realized version of itself.

All that said, this new non-dualist account of Athenian social being should again be seen as a very preliminary demonstration of the potential historical benefits of taking an ontological turn in our practice. Again, this account aims to be suggestive rather than exhaustive, and it may well beg a few further questions. To conclude Part III, we can address three possible issues up front.

The Mess of "Real Life"

First and most immediate, it might be pointed out that the behavior of actual flesh-and-blood Athenians did not always conform neatly to the prescriptions of the world-making common sense outlined in the preceding chapters. If the prem-ises of Athenian social being were as described above, then why were these prem-ises sometimes defied, even flatly contradicted in practice? Where in this account is there room to accommodate the sheer *messiness* of "real life"?

It is of course entirely possible to find examples of Athenians who appear to have acted against the prevailing ontological grain. In practice, families could be split by fiercely contested inheritance disputes. Some individuals could mistreat their wives, husbands, children, or parents. Others clearly did seek to minimize or avoid the military and/or financial obligations that came with membership of the *polis*.[1] Moreover, some prominent "advisors" of Demos doubtless used the mechanisms of the *politeia* for their own self-serving personal ends. No doubt too, the conduct of some others was affected by personal enmities with rivals, even when this conduct might potentially threaten the well-being of the social body as a whole.[2]

More serious, the very unity and integrity of this social body was on two occasions ruptured in the late fifth century, when elite constituencies assumed responsibility for managing the *polis* in 411 and 404, briefly supplanting *demokratia* itself. And perhaps more serious still were a number of attested violations of the ritual relations between Demos, land, and gods. The most notorious of these would include: the repeated desecrations of *ta hiera* in Athens during a time of plague (430–427 BC); the mutilation of many statues of Hermes and the profanation of the rites of the Eleusinian Mysteries at elite drinking parties shortly before the disastrous Sicilian campaign (415–413); and the failure of commanders to retrieve the bodies of casualties for burial in Athens after the naval battle of Arginusae (406).[3]

But there are at least two good reasons to believe that these various actions were neither commonplace nor acceptable at the time.

First, most of our evidence for questionable conduct by individuals comes from speeches delivered during legal proceedings. And one should hesitate before taking these sources as straightforward accounts of typical, representative behavior in the *polis*. Fewer than one hundred such speeches survive in their entirety. Together they allege transgressive acts by perhaps several hundred named individuals spread over the course of roughly one century, from ca. 420–320 BC, a tiny proportion of the social body. Moreover, in the majority of cases, the alleged transgressors were likely or demonstrably members of the social body's wealthier *oikoi*, Athenians whose conduct cannot automatically be taken to exemplify "normal" behavior within the *polis* as a whole. Furthermore, given the distinctly non-modern conditions in which disputes were staged, where cases were determined by large panels of "civilians" without help from legal experts, witness cross-examinations, or strict rules concerning the admissibility of evidence, there was little to deter litigants from exaggerating if not outright inventing the sundry wrong-doings of their opponents. So this kind of testimony cannot always be taken at face value, to say the very least. Then again, even if every single one of these alleged violations of laws and norms really did happen as described, this would still not necessarily rule out the account of social being presented in the preceding chapters. After all, it was precisely because such actions were deemed problematic that they were cited in forensic speeches in the first place. By definition, they were the exceptions that proved the rules.

Second, the various communal transgressions mentioned above all derive from the anomalous period when the *polis* was under the often extreme stresses of the Peloponnesian War. Indeed, the most serious of these actions occurred at moments when it could be said that the social body was experiencing war-induced threats to its very existence, circumstances that were therefore, by definition, extraordinary.

Thus, the most egregious of the violations of the traditional *hiera* of the *polis* occurred when a plague fell upon Athens while the entire population of Attica was sheltering within its walls for protection from annual Spartan invasions. As Thucydides so memorably describes, in the resulting overcrowded, refugee-like conditions, the spread of the contagion was seriously exacerbated, dead bodies piled up unburied in sanctuaries, and a general state of lawlessness prevailed among a demoralized people. Likewise, both of the occasions when *demokratia* was suspended came in the aftermath of catastrophic military defeats suffered by the *polis*. The regime of the Four Hundred was installed in 411 in a time of extreme duress, when Athens had just lost thousands of men and most of its navy during the disastrous campaign in Sicily (415–413), when the Spartans had established a permanent fortified position about fourteen miles north of the city at Decelea, allowing them to control all of Attica beyond Athens and Piraeus, and when many "allies" of the *polis* were actively seeking to defect from the Athenian empire. As for the far harsher, more destructive regime of the Thirty that held sway in Athens from the spring of 404 to the fall of 403, this was effectively imposed upon the Athenians by the Spartans and their allies as one of the conditions of the surrender which ended the Peloponnesian War. Even then, in neither case was the social body of Demos ever completely dissolved, since critical masses of Athenians who still claimed to embody this corporate persona actively defied both regimes, ultimately helping to secure the restoration of full *demokratia* not too long afterwards.

So it would be mistaken to suppose that the kind of recursive analysis pursued in the preceding chapters would have any more trouble accommodating the above transgressions than a more conventional historicist account. Indeed, by premising its account of the Athenian *politeia* on the world-making common sense of the Athenians themselves, this alternative mode of analysis can help us understand a little more precisely and immediately how and why such aberrant behavior might have seemed and felt so problematic at the time. It can help us to appreciate more acutely why, say, those who squandered the resources of their *oikoi* deserved forms of *atimia* for harming the *polis* as a whole, why being caught in the act of seducing the wife of a fellow-Athenian could merit a summary death penalty, and why those who sought to introduce new gods or stole from sanctuaries could incur the same ultimate punishment. It can help us to appreciate that much more keenly the potentially cosmic consequences of filling sanctuaries of the gods with unburied, plague-infected human corpses, mutilating statues of Hermes, profaning the Eleusinian Mysteries, or failing to retrieve the bodies of the war dead for burial in

Athens. It can help us to explain what exactly was at stake when Athenians were authorized to kill those who sought the "dissolution" (*katalusis*) of Demos. And it helps us to apprehend at an appropriately ontological level what a "dissolution of Demos" would actually have involved in the first place.

More generally, one should stress that common sense ontological and metaphysical commitments, however ubiquitously and unquestioningly held, are not ultimately capable of ruling out the possibility of their own transgression. Even when such commitments are deeply embedded in the fabrics of routine experience, at once presupposed and reproduced by all the most basic and essential practices of everyday life, they do not in the end function as inviolable iron laws of existence. One feels confident in asserting that there has never been a historical world where conduct conformed spontaneously and invariably to the a priori prescriptions of its prevailing template of social being. There will always be "gaps" between the world-in-theory and the world-in-practice.

For example, one does not have to be a Foucault, a Latour, or a Haraway, to see that modern western experience does not always compartmentalize itself neatly into discrete binary realms, like the human and the non-human, the sacred and the secular, state and society, public and private. And this is to say nothing of modernity's all too many "messy" contradictions, whereby the pursuit of a free and equal human individuality by westerners has directly or indirectly entailed the systematic exploitation of millions of indigenous peoples, colonized subjects, slaves, and industrial helots, along with the systematic denial of human rights to women, minority groups, and historically disadvantaged populations.

Yet it also seems fair to conjecture that future historians would still have great difficulty making meaningful sense of our current mode of life if they knew nothing of our peculiar world-making common sense, which objectifies social being as a disenchanted, functionally differentiated terrain inhabited by naturally free and equal human atoms. To be sure, familiarity with this general model of experience will not explain per se every single eventuality that has arisen in western polities since the eighteenth century. But without such knowledge, the most essential of the political, legal, social, economic, and other mechanisms that sustain our lives, that define our characteristically modern way of being human, would be all but unintelligible.

So it is with the account of the Athenian *politeia* that is presented above. This account does not pretend to cover all known or possible eventualities in the *polis*. Doubtless, the Athenians did not always live up to the expectations of their *politeia*. And doubtless, their world was not innocent of its own internal contradictions. But compared to those which have confounded our own modern polities, the Athenian contradictions would have been significantly less egregious, as an ontological history of the *polis* would help us to see. For the record, probably the most serious of them arose from the fact that many life-sustaining mechanisms of *demokratia* could not function effectively without the leadership and other personal resources that were provided by wealthy Athenians. And here, the resulting tensions would not have been between the sectional interests of rich and poor, as our own experience of egalitarian democracies might predispose us to believe. Rather, they would have been tensions between the basic principle of continuous self-rule by a unitary social body and the special, elevated status that was almost inevitably enjoyed by individual members of that body when they made extraordinary personal contributions to its general well-being.

Other Voices

But for some readers, there may be a more serious issue with the book's recursive account of the Athenian *politeia*, namely that the world it describes seems to depart at times quite significantly from the world that is presented to us in the texts of certain well-known Greek authors, Athenian and otherwise. And this objection might appear to derive greater force from the fact that some if not all of the authors in question are today revered as luminaries of "Greek thought," as if their ideas were somehow representative of "the Greeks" in general. Indeed, not a few of the thinkers in this category, perhaps especially those whose ideas about the world most seem to anticipate or resemble our own, would be considered among the early progenitors of the entire western tradition of intellectual enquiry.

One thinks here, for example, of the very different accounts of *demokratia* that one finds in works by Plato and other elites, who quite explicitly equate the *demos* in question with its relatively uneducated majority. In its more extreme forms, this critique manifests itself in the claim that the Athenian *politeia* was little more than the self-interested rule of "the poor," "the weak," or "the ignorant" over the richer, stronger, more educated minority.[4] Thus, a brief tract by an anonymous author, conventionally known as the "Old Oligarch," caricatures Athenian *demokratia* as a systematic attempt by "the worthless" (*hoi poneroi*) to exert power over "the worthy" (*hoi khrestoi*).[5] And then there is the notorious argument made in Plato's *Gorgias*, where the figure of Callicles expresses the claim that *demokratia* is not just undesirable but positively unnatural, since it authorizes "the weak" to rule in their own interest over "the strong," who are the "natural" rulers of any *polis*.[6]

In short, one can quite readily find voices in classical Athens that were willing to question common sense presuppositions about the very nature and purpose of *demokratia*. And some were even willing to challenge the self-understanding of Demos itself, the human essence of the *polis*, defining it merely as the self-realization of the "poor" majority rather than as the perpetual corporate persona of all Athenians. What then of the other foundations of the cosmic ecology in Attica? Were they too subject to critique?

The case of the Athenians' consanguineous, familial relations with their fatherland/motherland is rather less straightforward. To my knowledge, no classical author ever directly questions the prevailing truth that the Athenians were the original, indigenous inhabitants of Attica. To be sure, not all of the sources who mention this autochthony acknowledge the specific conviction that the first Athenians were literally "earth-born," though mere failure to mention this detail cannot automatically be taken to imply an express denial thereof. It is nonetheless possible to find texts which might challenge this particular article of faith. For example, like other tropes of Athenian funeral orations, the earth-born origins of Demos is treated with an unmistakable irreverence in the speech credited to Pericles' consort Aspasia in Plato's *Menexenus*.[7] And in Plato's *Republic*, Socrates explicitly cites a story about a people born from the soil of its homeland as an example of the kind of "noble fiction" (*pseudos gennaion*) that is typically invented to legitimize an established *polis* order.[8]

Moving on to the third of the foundations of Athenian social being, one should stress up front that it would be surprising indeed, shocking even, if significant numbers of Athenians questioned the prevailing common sense about the gods. For reasons discussed earlier, the reality of gods in this particular metaphysical conjuncture was not just a self-evident truth, but the most elemental truth of existence. That said, it is certainly possible to find texts which question conventional understandings of the precise nature of divinity and divine agency.

By the later fifth century, there was already a well-established intellectual tradition of critiquing the traditional Homeric vision of anthropomorphic gods.[9] Among the first to mount such a critique was Xenophanes of Colophon (active ca. 520 BC), who seems to have supported a more abstract and somehow more unitary conception of divinity, castigating his fellow Greeks for imagining that gods could possibly be prone to the same limitations and failings as mortals.[10] Elsewhere, Heraclitus of Ephesus (active ca. 500 BC) represents divinity as a kind of ultimate cosmogonic principle, whereby the universe is held in a condition of balanced unity through dynamic tensions between opposites.[11] Pericles' friend Anaxagoras of Clazomenae (active ca. 450 BC) apparently associated divinity with a kind of transcendent, autonomous "Mind" (*Nous*), an essentially immaterial force that had progressively fashioned the world of experience, with its myriad distinct phenomena, from a primordial mass of unformed matter.[12] And perhaps more inflammatory still, the treatise *On the Gods* by the celebrated sophist Protagoras of Abdera (ca. 490–420 BC) is said to have opened with the following declaration:

Concerning the gods, I am unable to discover whether they exist or not, or what they are like in form; for there are many hindrances to knowledge, [including] both the obscurity of the subject and the brevity of human life.[13]

But it was one thing to question conventional ideas about the nature or the knowability of divinity. It was quite another to deny divinity's existence altogether. While a number of the intellectuals mentioned above may have been accused of being "godless" (*atheos*) at some point, the evidence hardly permits certainty about the fairness of such charges. The one figure who appears to have been almost universally regarded as godless at the time was a poet called Diagoras of Melos (later fifth century BC). But little is known of his actual ideas beyond what is found in anecdotal reminiscences recorded by much later authors.[14] Otherwise, among extant classical texts, the most explicit claims to belief in a godless cosmos are to be found in fragments from two tragedies. However, given that the claims in question are put in the mouths of two characters, Bellerophon and Sisyphus, who would in the end suffer proverbially devastating divine punishments, these texts at most attest only to the imaginability of godlessness, not to its contemporary currency.[15]

In sum, while few if any Greeks in the classical era demonstrably believed in a godless universe, a very small number did propagate unconventional ideas about the essential nature and knowability of divinity. So how then would a non-dualist ontological history of Athens accommodate all of the various "other voices" discussed in this section?

It would begin by stressing that such heterodox ideas were produced, disseminated, and actively embraced only by relatively tiny, highly educated elite minorities. As far as one can tell, they were not at all representative of common sense "Athenian thought," the social knowledge that informed the actual life-sustaining practices of most Athenians most of the time. Indeed, all of the claims discussed above were self-consciously oppositional. Formulated by renegade intellectuals, they were published precisely because they challenged the prevailing common sense knowledge, the knowledge which did not have to be written down. They aimed expressly to question and/or subvert established presuppositions about gods, land, and people and the relations between them, the kinds of presuppositions which sustained the Athenian *politeia* as whole. And for this reason, a non-dualist ontological history of Athens, unlike a conventional dualist account, would insist on observing a crucial analytical distinction between these heterodox and orthodox claims about the essential contents of experience, a distinction that was noted earlier in Chapter 9.

For when seen from a conventional dualist perspective, Athens merely harbored a range of competing, epistemologically equivalent accounts of an objectively existing world, a diversity of ideas, beliefs, ideologies, discourses, and so forth. Some of these accounts may have been "orthodox" or "dominant," while others may have challenged or opposed the prevailing ideas. But all of them, whether dominant or oppositional, were still in the end purely ideational constructs. They were bodies of thought that were ontologically distinct from the mind-independent world which they aimed to render intelligible. Thus, we historians can ultimately determine for ourselves the truth status of such accounts, since our modern science equips us better than the ancients to assess the correspondence of their thought to any objective reality.

However, when seen from the alternative non-dualist perspective proposed in this book, the picture changes quite decisively, not least because there would no longer be a universal, mind-independent world out there against which to measure the truth status of any ancient claims to knowledge. The orthodox claims discussed in the preceding chapters would be part of the world-making common sense of the age, the prevailing a priori knowledge that underpinned all of the practices and mechanisms which sustained Athenian social being. They would thus express the shared ontological and metaphysical commitments that were woven into the very fabrics of lived experience, commitments that actively helped to make the world of the Athenian *politeia* whatever it really was in the classical era. By contrast, the various oppositional claims surveyed above would amount merely to heterodox ways of seeing the particular world thus realized, to an assortment of purely ideational a posteriori worldviews. These worldviews of a renegade minority would still presuppose broadly the same non-modern, divinely ordered metaphysical conjuncture as the prevailing social knowledge. They would just see some of the a priori contents of that world, like divinity itself and *demokratia*, through somewhat different eyes.

In other words, for the purposes of an ontological history, conventional or "traditional" claims about the givens of existence would be far more significant than the more intellectually impressive or sophisticated counter-claims made by the likes of Plato, Protagoras, and Anaxagoras. For unlike their orthodox counterparts, the oppositional claims would not actively participate in constituting reality as it was commonly known and lived at the time, and thus would not correspond precisely to that reality. So when examined through the lens of a non-dualist historicism, the testimony of our "other voices" does not in fact threaten the account of Athenian social being that is presented in the preceding chapters after all. If anything, these dissenting voices indirectly support that account by helping us to confirm what was and was not the prevailing world-making common sense of the time.

To substantiate the point further and sharpen this analytical distinction between world-making common sense and worldview, we might very briefly consider the case of the *History of the Peloponnesian War* by Thucydides the Athenian.

For many readers today, this particular classical text, perhaps more than any other, seems to speak to us in a distinctly proto-modern or modern voice. In the eyes of historians and political scientists alike, the *History* has long appeared to take for granted something like a modern secular, anthropocentric universe, a disenchanted world where humans are the only meaningful actors and agents, where actions are determined by a sometimes brutally "rational" pursuit of power or self-interest, and where a prevailing materialist truth standard allows a clear distinction to be drawn between mythical fiction and objective history. In short, the *History* seems to portend a world that is altogether amenable to the more "scientific" historical practice with which its author continues to be associated.[16]

But for some time now, there has been a steadily growing resistance to this image of the *History* as a kind of harbinger of modern-style historical and political analysis. Ever since the first publication of Francis Cornford's *Thucydides Mythistoricus* in 1907, classicists have readily identified the traces of story patterns in the *History* that resemble those found in earlier myths and tragedies.[17] Many have also noted affinities of language, thought, and imagery between Thucydides and more obviously "traditional" authors, such as Homer, Herodotus, and even Pindar.[18] And as Moses Finley reminds us, there was nothing startlingly innovative or even unusual about Thucydides' characteristic concern with the sources, means, and ends of human forms of power:

> It must be admitted that Thucydides was not an original thinker. The general ideas with which he was obsessed were few and simple. He had a pessimistic view of human nature and therefore of politics. Some individuals and some communities, by their moral qualities, are entitled to positions of leadership and power. But power is dangerous and corrupting, and in the wrong hands it quickly leads to immoral behavior, and then to civil strife, unjust war and destruction. These were familiar themes among poets and philosophers.[19]

Nor for that matter is the *History* self-evidently informed by any modern-style form of historical consciousness. While the text may at times question the precision of the information conveyed by the likes of earlier poets and Herodotus, stressing its own dependence on authorial autopsy and human eye-witnesses, it plainly does not recognize anything like our basic categorical distinction between "myth" and "history."[20] As far as one can tell, figures like Deucalion and Hellen, the Cyclopes and the Laestrygonians, Minos, Pelops, Heracles and Theseus, were all just as historically real to Thucydides as Pericles and Cleon.[21] Indeed, our modern sense of a definitive difference or distance between past and present, our historicist sense of the past's essential alterity, seems to be entirely missing from the *History*. As Zachary Schiffman, a specialist in early modern historiography, has emphasized, the ancients were wholly unacquainted with our modern idea of "anachronism," not least because this particular idea was only invented in the Renaissance era.[22] As a result, "when Thucydides looked back in time, he perceived . . . a past inextricably bound up with the present, a past-made-present."[23]

Consider, for example, his account of the synoecism of Attica. For when he describes how Theseus once united multiple primeval *poleis* in the region to create the single *polis* of "the Athenians," Thucydides takes it entirely for granted not only that the mechanisms which sustained the now unitary *polis*, like the new pan-Attic council, were more or less identical to those which underpinned the Athenian *politeia* in his own much later time, but also that all the predecessor ur-*poleis* had been hitherto organized along similar lines.[24] And of course, it was precisely this sense of an essential identity and continuity between past and present that allowed the historian to presume that his account of one particular Greek war might eventually turn out to be a "possession for all time."[25]

In short, as Nicole Loraux once put it, "Thucydides is not a colleague" of the modern historian.[26] For all its remarkable rigor and precision, the *History* was in the end conditioned by a distinctly non-modern form of historical consciousness, one that was not significantly different from that which animated epic poetry, tragedy, the *Histories* of Herodotus, and all the commemorative speeches, ceremonies, and monuments of the classical Athenians. And perhaps this is none too surprising. For I would also argue that Thucydides' account is sustained by metaphysical and ontological commitments that were also broadly conventional for the time. The *History* assumes a basic template of social being, a model of "the world," which was essentially the same as that which was both presupposed and reproduced by the Athenian *politeia*.

Thus, Thucydides takes it for granted that the human constituents of *poleis* acted in and upon the world as unitary social bodies. Of course, like any observant Greek of the age, he could see that differences of background, interest, and opinion might at times divide the particular members of these bodies, and that individual leaders could have a far-reaching influence upon particular deliberative outcomes.[27] But even in the case of the fractious, restless Athenians, he gives us no reason to doubt that their Demos was still a discrete corporate person in its own right, one that ontologically preceded and outlived the particular individual flesh-and-blood persons who happened to embody it any given time. Hence, when Thucydides portrays Athenians giving what are arguably the two most programmatic, manifesto-like statements of the basic imperialist rationale which guided the conduct of Demos before and during the war, he does not tell us which particular speakers were responsible for the words in question. In both cases, the speakers are merely anonymous "Athenians." And presumably he doesn't tell us their names because their names did not matter as such, since their words conveyed the ideas of an entirely different person, the corporate person of Demos, which these speakers were effectively embodying at the time.[28]

Similarly, Thucydides never once questions the shared conviction that the Athenians were a uniquely autochthonous people who thereby enjoyed a special relationship with their homeland. True, he makes no direct reference to stories of humans born miraculously from the soil of Mother Attica. But he readily accepts the historical reality of the early kings Cecrops and Erechtheus about whom such stories were told.[29] More important, he himself categorically endorses the standard Athenian claim that "the same people have always been living in the land of Attica" (*ten Attiken [gen] . . . oikoun hoi autoi aiei*).[30] And for good measure, the very same claim is later repeated in much the same terms by Pericles, the Athenian leader most obviously admired by Thucydides, near the beginning of his famous funeral oration:

> The same people has always inhabited this land of ours (*hoi autoi aiei oikountes*) from generation to generation right up to the present day. And they, by their courage and their virtues, have handed it down to us a free land.[31]

Finally, there is the rather less straightforward matter of the place of the gods in the world of Thucydides' *History*. Of course, divinities are not present as actors in the text in the ways that they are in earlier Greek works. But this does not necessarily mean that the world presumed by the *History* was essentially godless or secular.[32] Given a conjuncture where gods self-evidently administered the very conditions of existence, it seems safe enough to assume that a classical text will share the basic metaphysical commitment of the time to some kind of divinely ordered cosmos unless it furnishes incontrovertible evidence to the contrary, which the *History* manifestly does not do.

Indeed, the *History* is fairly loaded with references, direct and indirect, to divine-human relations.[33] True, unlike so many other Greek works, it does not presume to know what those relations might have looked like from the divine side, from a god's-eye perspective. A work which claims to derive most if not all of its information from authorial autopsy and human eye-witnesses could hardly pretend otherwise. It is also true that the text is more than willing to expose those who were guilty of abusing ritual activities for their own immediate self-serving purposes.[34] But at no point does it give us cause to question the meaning or value of such activities in general, dismissing them all, say, as crude exercises in popular manipulation or as pure folly, because there were really no gods there in the cosmos in the first place. In other words, the *History* makes no claims to understand the precise designs, actions, or even natures of the gods of the Greeks, because such phenomena were not ultimately knowable or verifiable within the strict parameters of a Thucydidean historical practice. But there is nothing in the text that is wholly inconsistent with the prevailing presupposition that divinities were really present and active in the world of experience.

More positively, as a number of specialists have shown, the treatment of oracles in the *History* broadly aligns with "mainstream" views of the time.[35] As far as one can tell, the author accepts both the characteristic ambiguity of oracular pronouncements and the essential truth of the information that they conveyed. The text may at times be quite harshly critical of human efforts to interpret and act upon certain specific oracles, but at no point does it give us cause to question the overall utility or veracity of the oracles themselves.[36] And these general observations have now been strengthened by Lisa Kallet, who makes a compelling case that Thucydides' work quite explicitly recognizes the possibility of "divine intervention in human history," citing in support his account of the plague suffered by the Athenians in 430–427 BC. As Kallet demonstrates, the text quite unambiguously presents this event as assistance rendered to the enemies of Athens by Pythian Apollo, who had earlier promised a delegation of Spartans that he would actively help them to victory if they fought with all their might.[37]

To summarize, the *History* of Thucydides broadly shares the prevailing ontological and metaphysical commitments of its particular time and place. It takes for granted the very same general model of "the world" that was presumed by the Athenian *politeia*, even if its view of certain essential components of that world was in some ways unconventional. Thucydides's gods may not have been the relatively knowable, sometimes visible gods of Homer, Sophocles, or Herodotus. His Attica may not have been a primordial "mother" figure who miraculously gave birth to the first Athenians. And at least when under the sway of an influential leader like Pericles, his Demos of the Athenians may not always have been the entirely self-governing social body that it took itself to be. But if the worldview of the *History* is thus somewhat heterodox, the world that it viewed was essentially the same one that was seen and experienced by most if not all Athenians of the time. And this world of ancient Greek experience bore no resemblance whatsoever to the materialist, anthropocentrist, secularist, and individualist world of our own modernity. So far from contradicting the general picture of Athenian social being presented in the preceding chapters, the *History* if anything offers that account a measure of implicit support.

Tales We Can Tell

Finally, some readers may feel that the book's alternative account of the Athenian *politeia* is simply unrealistic, in that it treats the period 480–322 BC for the most part synchronically, as a single extended moment in time, thereby possibly minimizing or ignoring the impact of the many significant historical events and changes which occurred during those years.

Of course, one can hardly understate the consequences of the era's major historical developments. Most obviously, the fortunes of the *polis* were transformed by the rise of the Athenian empire and then again by its later fall. The Peloponnesian War had any number of far-reaching effects on social being, diminishing or outright destroying so many lives and livelihoods. And the very autonomy of Demos itself as an agency became steadily more compromised and constrained after the defeat by Philip II of Macedon at Chaeronea in 338 BC. Doubtless too, the period witnessed some notable changes to the Athenian *politeia* itself. For example, new divinities, like Artemis Aristoboule, Asclepius, and even the Thracian goddess Bendis, were introduced to the local pantheon. From the mid-fifth century on, lotteries were increasingly used to determine who should perform functionary services for the *polis*, and most of those services, including assembly attendance, ultimately came to be remunerated with modest stipends. And following the restoration of *demokratia* in 403, panels of *nomothetai* ("lawgivers") were selected from those serving as judges to decide whether newly proposed *nomoi* ("laws") should take their place among the permanent rules of the *polis*.

Yet for all the self-evident importance of the above developments and many others besides, one should not at this point have to defend a synchronic treatment of this or any other historical era. Innumerable social, cultural, and other histories have been written in this mode for decades, forming a tradition which extends back at least to ground-breaking studies like Marc Bloch's *Feudal Society* (1939) and E. M. W. Tillyard's *The Elizabethan World Picture* (1942).[38] More to the point, a synchronic approach would seem to be positively necessary for a work like this one, whose express aim is to recover the a priori essences and foundations of social being in a particular time and place. Since metaphysical and ontological givens of existence tend not to be prone to continual change by definition, it is hard to see how they could be adequately represented and analyzed in a more conventional narrative history format. Whatever the case, it is clear enough in the Athenian instance that the metaphysical and ontological givens did not change appreciably during the classical period.

For example, Athenian engagements with divinity were no less frequent and no less meticulous in 320 than they had been back in 480. In all likelihood, there were more gods in the Attic pantheon at the end of the classical era than there had been at its beginning. Likewise, for all the devastation wrought upon the land and farms of the *polis* during the Peloponnesian War, the period saw no discernible change in the interdependent relationship between Mother Attica and her autochthonous progeny. Nor was there any fundamental change in the animating logics and principles of the Athenian *politeia*. From 508 until 320, and for many years thereafter, *demokratia* in Athens meant self-management by a perpetual, unitary social body, whether the members of this body were acting severally as its parts or in concert as the whole. And throughout the classical era, this *politeia* remained the human contribution to a cosmic ecology of gods, land, and people in Attica. While innovations were surely necessary to maintain the alignment between life-sustaining practices and the prevailing model of social being as historical circumstances shifted and evolved, the model itself remained essentially unchanged.

Indeed, one might further suggest that a non-dualist historicism would encourage us to consider afresh the "realism" of conventional diachronic accounts of Athenian experience, prompting a range of legitimate questions. For example, how far might narratives which are organized around modern chronometric devices, like the Gregorian calendar and fixed historical "periods," misrepresent or distort the experiences of a non-modern people who objectified and measured the passage of time entirely otherwise? Given a world which lacked our historicist notion of difference or distance between past and present, do our standard stories about Athens impose on events a modern sense of forward motion or progress through a homogeneous, empty time which seems to have been entirely missing in Athens itself?[39] How far should we temper, say, our tendency to see *demokratia* as a "revolutionary" or "progressive" enterprise when it was experienced as the more or less timeless order of things by the classical Athenians? More generally, given a world which presupposed a radically different way of being human, who should determine the appropriate contents of our narratives in the first place, our subjects or ourselves? Who in the end should decide what constituted a significant "change," "event," or "development," one that would be eminently worthy of record?

More important, an ontological turn in our discipline would itself raise the prospect of a new kind of diachronic history, a new horizon of tales we can tell. These new narratives would be stories about the evolution of different ways of being human, stories that expressly explore how changes in particular metaphysical conjunctures conditioned the formation of new subjects, objects, relations, and processes at the ontological level. Thus, even if one lacks all the requisite evidence to tell such a tale, one could certainly imagine an alternative Athenian history along these lines, a narrative that would seek to recover tectonic shifts in the fabrics of being over the *longue durée*, from what we call the neolithic era (ca. 7000–3000 BC) down to the classical era and beyond.

This grand narrative would no longer be a story about multiple periods in the lifetime of a single, more or less continuous world. In all likelihood, it would be a tale about a succession of at least three distinctly different worlds, all with their own particular ways of being human. It would begin with the "pre-Greek" world of those who first settled in the area of the Athenian acropolis, before the coming of the Indo-Europeans in ca. 2100 BC. It would then include the world which culminated in the formation of a near-eastern-style order in the later centuries of the second millennium BC, with its palace complex on the citadel, its centrally managed mode of life, its Indo-European gods and language, and its ruling *wanax*, who most probably served as some kind of intermediary between communities human and divine. And of course it would also include the later world of the Athenian *polis*, the world of this book, where, ultimately, the life of a unitary social body was sustained by a primordial cosmic ecology of gods, land, and people.

Whether or not it is actually writable, this alternative non-dualist history would be quite radically different from the kind of grand narrative with which we began in Chapter 1. It would not be a standard story of changing political, social, and economic practices on the one hand and changing beliefs, values, and ideologies on the other. It would instead be a tale about changes in the a priori contents of non-modern experiences, changes in world-making common sense. It would be a story of shifts in, say, prevailing forms of subjectivity and sociality, in modes of shared memory and rationality, in the nature and identity of the super-human forces and agencies with which humans coexisted, and in the very sources, means, and ends of life itself. And unlike conventional exercises in cultural history or *Begriffsgeschichte*, this diachronic ontological history would explore how such world-making essences were effectively realized as phenomena in everyday experience. It would seek to show how they came to be both presupposed and reproduced by narratives, ceremonies, monuments, images, and the most urgent life-sustaining practices, such that they eventually acquired the truth of their own self-evidence, as if they had all really been there all along.

Conclusion

NEW HORIZONS OF HISTORY AND CRITIQUE

And what should they of England know who only
England know?

RUDYARD KIPLING

FOR ALL THE undoubted achievements of our conventional historicist practice, we should still be troubled by some of the histories that it is liable to produce. For in order to render the experiences of past peoples mutually intelligible and commensurable with our own, it takes for granted a uniquely modern model of social being, one that would ontologically re-engineer all non-modern lifeworlds to comply with modernity's peculiar dualist metaphysical conditions. It thus effectively dismantles the complex materio-cultural essences of those worlds, phenomena like gods, cosmic ecologies, corporate persons, and dividual selves, sundering what we take to be objective materialities from subjective idealities. It then refashions those materialities and idealities into more familiar phenomena, like states and societies, economies and individuals, cultures and ideologies, all of which were of course potentially realizable in the modern-style metaphysical conjuncture that we presume prevailed at the time. And it then asks how our quasi-modern subjects might once have produced and sustained these quasi-modern worlds, these worlds of our own making. The inevitable results of this practice are "as if" or "what if?" histories, bold thought experiments that posthumously homogenize non-modern ways of being human, figuring them all as imperfect premonitions of our own.

As I hope this study has shown, there is another possible practice. Rooted in a non-dualist, quantum-style historicism, this alternative paradigm obliges us to recognize the metaphysical autonomy and integrity of extinct lifeworlds, and thus to account for the many ontological variabilities of the past. For all practical purposes, this means relinquishing our god-like prerogative to refashion past worlds in the image of our own. It means restoring to non-modern peoples a certain power to determine the ultimate truths of their own existence. It means studying those peoples recursively, on their own terms, guided by their world-making common sense, not ours. And if we are prepared to abide by these new rules of historical engagement, we can produce histories that are at once less ethically troubling, more philosophically robust, and more historically meaningful.

If so, this proposed ontological turn in our practice clearly has potential ramifications that extend far beyond the field of Greek history. To conclude the book, I would like to suggest what seem to me to be four of the more significant of these implications.

First and most immediately, the new paradigm would oblige us to adjust our conception of the past in some fundamental ways. Perhaps above all, it would require us to suspend that sense of transhistorical metaphysical continuity which allows us to imagine that all past worlds belong ultimately to a single common spatio-temporal dispensation. It would require us to resist thinking of past peoples as participants in a single, universal human story, one that progresses inexorably through homogeneous, empty time towards a predetermined telos in capitalist modernity. And while it would not necessarily force us to give up the convenience of our standard chronometric devices, it would nonetheless oblige us somehow to suppress the peculiarly modern species of historical consciousness which has produced those devices. For if we are to take seriously the possibility of many different ways of being human in many different worlds, as a non-dualist historicism urges us to do, we must somehow reimagine the past as a site of metaphysical multiplicity, diversity, and variability, as a plurality of lived presents.

Second, embracing the new paradigm would in turn mean rethinking the relations between past and present, because it would mean coming to terms with the sheer metaphysical anomalousness of our own modernity. It would mean adjusting our historical common sense to accommodate that great divide which separates the modern from all non-modern ways of being human. And this adjustment would itself have certain practical consequences.

Perhaps most obviously, it would cause us to question histories which take a measure of metaphysical continuity between worlds past and present for granted. It would complicate our sense of ontological commonalities across time and space, deterring any easy assumption that the subjects, objects, relations, and processes produced by one historical world can be readily realized under the metaphysical conditions of another. And it would expressly resist efforts to weave together stories about uniquely modern phenomena, like states and economies, from bodies of non-modern evidence. More generally, it would encourage us to rethink our conventional geo-chronological orderings of past experiences. In particular, by stirring us to recognize the profound differences which separate modern from non-modern European lifeworlds, it would destabilize not only our conventional ancient-medieval-modern stadial sequences but also the routine, common sense distinctions that we draw between the experiences of "western" and "non-western" peoples.

Third, a commitment among historians to an alternative, non-dualist historicism could in turn have reverberations in the wider intellectual landscape, not least by adding significant new weight to existing critiques of the disciplinary divides which have hitherto structured the production of modern knowledge.

For a start, committing to the cause of studying past peoples on their own ontological terms, in their own metaphysical conjunctures, would further imperil the already tenuous, somewhat ethically dubious division of labor between history and anthropology, the two disciplines that have traditionally specialized in the analysis of non-modern experiences. As noted earlier in the book, this same general cause has already been embraced by a number of leading cultural anthropologists, whose work quite explicitly challenges the very notion of an abstract, universal species-subject called *anthropos*, their discipline's defining object of study. At the same time, if we are prepared to liberate all non-modern peoples analytically from our modern historicist dispensation of space-time, as I would recommend we do, it would no longer be at all clear which of those non-modern peoples would "belong" to history and which to anthropology. And as this book's Athenian case study quite clearly illustrates, even if historians were to formulate their own distinctive procedures for analyzing non-modern ontologies, it would still be far from self-evident where the work of "history" would end and that of "ethnography" might begin.

Pursuing this line of thought further, one could then envisage how the fruits of a non-dualist history and a non-dualist anthropology might together add further momentum to the growing calls in various quarters for a genuinely postdisciplinary intellectual environment. As we saw earlier in Chapter 8, a number of different bodies of critical theory have seriously questioned the a priori, mindindependent self-evidence of the sundry phenomena which have hitherto defined the particular parameters and/or objects of study in each discipline. By drawing attention to the ontological and metaphysical discontinuities that separate modern from non-modern worlds, historians and anthropologists could add significant weight to these arguments. They could actively encourage non-specialists to see that our taken-for-granted dichotomies in experience between the material and ideational, the natural and cultural, the animate and the inanimate, the human and non-human, the sacred and the secular, the political and the social, and so forth, are not timeless givens of existence that can prevail in each and every historical world. They could help others to see objects like state, society, economy, and religion as historically contingent, quantum-like effects, as phenomena that are ultimately realizable only under the metaphysical conditions of our capitalist modernity. And in so doing, they might hopefully help those others to recognize that our entire apparatus of mainstream knowledge production is thus, as it were, pre-programmed to produce and reproduce the world-making common sense of just one way of being human.

Fourth and finally, even if an ontological turn is ultimately pursued by only small minorities of historians and anthropologists, the results could still contribute significantly to the general cause of critical theory itself, to the production of a more historically and ethnographically grounded counter-knowledge of own modern conjuncture. As was emphasized back in Chapter 9, recognizing the profound metaphysical discontinuities which separate the non-modern from the modern does not necessarily mean valorizing the latter at the expense of the former. On the contrary, by divesting our capitalist modernity of its status as the naturally ordained telos of our species story, an ontological turn in our practice would raise the possibility of a new strain of critical theory, something like a non-modern critique of the modern.[1]

Of course, one can readily concede that modernity, with its historically unprecedented levels of material prosperity, social complexity, and technological sophistication, has achieved significant progress when measured by its own peculiar standards. But a non-dualist historicism, one that would implicitly denaturalize and deuniversalize those same standards, could also help us to evaluate the extraordinary costs of any such achievement from a certain "outsider" perspective.

For example, if such help were needed, it might help us to see more clearly that the much-vaunted benefits of individualism have been enjoyed disproportionately by affluent minorities in the United States and western Europe, not by humanity as a whole. It might help us to see that these benefits have too often been accrued through the dehumanizing, soul-numbing toils of legions of industrial laborers, slaves, indentured servants, colonized peoples, sweat shop workers, and other producers the world over, beings whose claims to a meaningfully free and equal individuality have been all too easily minimized or overlooked. And it might help us to see how the pursuit of an allegedly universal imperative of self-improvement by capitalism's private leviathans, from the Royal African and East India Companies to Monsanto and Royal Dutch Shell, has forever corrupted, undermined, or outright annihilated innumerable local, sometimes centuries-old ecologies the world over, alienating multitudes from their ancestral certainties and their former selves.

For that matter, it can help us to see why a mode of social being which regards itself as the true, authentic expression of our natural humanity might be capable of unleashing such ecological devastation in the first place, given that "enlightened" moderns can all too easily regard non-individualist "others" as living somehow contrary to nature, as beings who are somehow less than fully human. And it can help us to see the often irreversible damage that the uniquely modern, relentless, global pursuit of "growth" and "productivity" has caused and continues to cause to the earth's forests, soils, water bodies, species populations, and atmosphere in our anthropocene age. If capitalism is in effect our biological destiny as natural individuals, one has to wonder in the end why Nature has implanted in us a commitment to a "mode of subsistence" whose need for perpetual growth, *sine die*, would inevitably deplete, degrade, and eventually exhaust the rest of her creation.

We do not have to settle for this "progress." The ultimate aim of this book has been to suggest that the past has seen many distinct ways of being human, each one of them bounded and shaped by its own local, contingent conditions of existence. So if this project has anything valuable to contribute to a cause larger than academic history-making, it is to suggest that effective critique of our present conjuncture must accordingly begin at the ontological and metaphysical levels. It must begin by questioning how precisely we determine what is always already there in our experience. Blame for the kinds of problems listed above cannot be traced to any one particular phenomenon, whether capitalism, imperialism, or environmental degradation. The cause lies ultimately in modernity's whole way of being human, in the peculiarly materialist, anthropocentrist, secularist, and individualist conditions of its existence, conditions which have made phenomena like capitalism, imperialism, and environmental degradation seem somehow normal, natural, and even inevitable in the first place.

It follows that truly effective attempts to imagine an alternative, less self-defeating, more just, more sustainable way of being human must also begin at the ontological and metaphysical levels. Such efforts must fundamentally rethink that entire template of being through which we know experience, that body of world-making common sense which forever separates the physical from the metaphysical, the material from the cultural, the human from the natural, the political from the economic, state from society, and individual subjects from one another. Lest we think that such a wholesale reimagining of social being is beyond us, we need only revisit the formation of western modernity itself to see that it is not. If certain peoples in the seventeenth, eighteenth, and nineteenth centuries were able to erect a great metaphysical divide between themselves and their predecessors, to reconfigure the very fabrics of their existence and modify their everyday thought and practices accordingly, we can surely do so again. And if we need examples to guide us in our world-making projects of the future, to expand our horizons of what is thinkable and realizable, we can find them, I suggest, in history's many other worlds, in the myriad non-modern worlds of the present and the past.

Notes

INTRODUCTION

1. The term "lifeworld" is used throughout this study to refer to the bounded totality of any historically distinct "way of life." It serves as a more inclusive, more analytically neutral alternative to standard terms like "society" and "culture," each of which, as we shall see, presuppose a peculiarly modern western mode of social being.
2. Chakrabarty 2000, esp. 3–113.
3. I would not be the first historian of the non-modern "West" to draw on the resources of postcolonial critique. For example, some medievalists, notably Dagenais and Greer (2000) and Symes (2011), have claimed that modernity has used historicism to "colonize" the European Middle Ages as its subaltern "other." Elsewhere, Kostas Vlassopoulos (2007) has argued that mainstream Greek historians have reified and Europeanized "the Greek *polis*" as an ancestral precursor of the modern nation-state.
4. To my knowledge, Geoffrey Lloyd (e.g., 2011, 2012) is the only other specialist in ancient Greek studies who has previously pursued an explicit interest in matters ontological, though his approach still remains within the broad compass of cultural history.

CHAPTER 1

1. E.g., Pomeroy et al. 2008; Morris and Powell 2010.
2. Anderson 2007.
3. While the account that follows is inevitably something of a caricature, the primary purpose here is to highlight the common body of often unacknowledged theoretical presuppositions which underpin all mainstream analyses of the Athenian way of life.

4. Meier 1990, 5, 22–23. Since most regard them as essentially political entities, Greek *poleis* tend to be categorized as "city-states" or "citizen-states": e.g., Hansen 1998, 2006.

5. Egalitarianism as the defining commitment of Athenian *demokratia*, see e.g., Ober 1989; 1996, 18–31; Meier 1990, 29–52.

6. E.g., Isocrates 7.20; Lysias 26.5; Thucydides 2.37.2, 7.69.2.

7. Enrollment in demes: Whitehead 1986, 97–109. The ten tribes: e.g., Jones 1999, 151–94.

8. Overview of Athenian decision-making institutions and their operations: Sinclair 1988. Assembly meetings: Hansen 1987.

9. Nature and operations of the council of 500: Rhodes 1972.

10. Essential "structural" role played by elite speakers in assembly deliberations: Finley 1974; cf. Kallet-Marx 1994.

11. Overview of the Athenian "legal system": Bers and Lanni 2003.

12. Overview of different legal procedures in Athens: Todd 1993, 77–163. Logistical details: Aristotle, *Constitution of the Athenians* 63–69.

13. Duties of Athenian officials: Aristotle, *Constitution of the Athenians* 43–62.

14. How "civilians" were able to "police" Athens: Hunter 1994.

15. Athenian generals: e.g., Hamel 1998.

16. E.g., Van Wees 2004, 45–113 (cavalry and infantry); 199–231 (navy).

17. See pp. 34–38 below.

18. Family life in Greco-Roman antiquity: e.g., Rawson 2011; cf. Hunter 1994; Cox 1998; Patterson 1998; Roy 1999. Greek houses: Nevett 2001.

19. Athenian marriage as a "social process": Patterson 1998, esp. 108–14.

20. Modern "common sense" view of female experience in Greek antiquity: e.g., Blundell 1995; Pomeroy 1995.

21. E.g., Jones 1999. *Gene, orgeones, thiasoi*, and other specifically "religious" associations: Parker 1996, 284–342. Phratries: Lambert 1998.

22. Metics in Attica: e.g., Whitehead 1977; Kennedy 2014.

23. General treatments of slavery in the Greek world: e.g., Finley 1980, 1981, 97–195; Garlan 1988; Fisher 1998. Public slaves: Ismard 2017.

24. See pp. 39–42 below for further discussion. Reconstruction of "the ancient Greek economy": Bresson 2007, 2008.

25. In the fifth century, a majority (*hoi pleious*) of Athenians still lived in the "countryside" (*khora*) of Attica: Thucydides 2.16.1.

26. Greek agriculture and land-use: e.g., Gallant 1991; Sallares 1991; Burford 1993; Foxhall 2007. Athenian silver mines: Hopper 1968; Jones 1982.

27. Athenian division of labor: Harris 2002. Athenian banking and finance: e.g., Cohen 1993. Manufacturing: Acton 2014.

28. This very hypothetical picture of wealth distribution in Athens broadly follows the account of Davies (1981, 35–37), which includes a very helpful (and visually striking) graph.

29. Athenian public finance: Pritchard 2015. History and operations of Athenian treasuries: Samons 2000.

30. Definition and expenditures of the "liturgical class" in Athens: Davies 1981, esp. 9–37. Administration of liturgical obligations: e.g., Christ 2006, 143–204.

31. General treatments of "Greek religion": e.g., Burkert 1985; Easterling and Muir 1985; Parker 2011.

32. Athenian festivals and other ritual observances: e.g., Parke 1977; Mikalson 1983.

33. Coexistence of "the irrational" with "rationalism" in ancient Greece: Dodds 1951. Persistence among Greeks of "belief" in "myths": Veyne 1988.

34. Overview of the Panathenaia: Neils 1992. Over-representation of elites in the Panathenaic procession: Maurizio 1998. Control of the cult of Athena Polias by the Eteoboutadai: e.g., Parke 1977, 17–18. Continuing control of at least some cults of the ten tribal eponymous heroes by traditional "aristocratic" families: Anderson 2003, 130–31. The Eleusinian Mysteries and the roles played therein by Eumolpidai and Kerykes *gene*: e.g., Parke 1977, 55–72.

35. When necessary, *polis* authorities would consult freelance professional "seers" (*manteis*) and "oracle-collectors" (*khresmologoi*): Parker 2005, 92–94.

36. *Demos* as the ultimate authority on religious matters in Athens: Parker 2005, esp. 89–115.

37. Pioneering examples in ancient Greek studies would include: Humphreys 1978, 1983; Vernant 1980; Loraux 1986; Vidal-Naquet 1986.

38. A selective sample of some of the more influential items in this diverse and ever-growing body of work on Athens and Greece in general might include: Ober 1989; Thomas 1989; E. Hall 1989; D. Cohen 1991; Zeitlin and Winkler 1991; Boegehold and Scafuro 1994; J. Hall 1997; Dougherty and Kurke 1998; Allen 2000; Morris 2000; Alcock 2002.

39. E.g., Halperin 1990, 88–112 (constructions of the body); Morris 1992, 103–55 (fifth-century burial practices); Hoepfner and Schwandner 1994 (urban grid-plans in classical settlements like Piraeus); Hanson 1995, 1996 (phalanx); Strauss 1996 (navy); Csapo and Miller 1998 (temporality); Balot 2014 (courage).

CHAPTER 2

1. If anyone remains to be persuaded of this, a succinct and compelling case can be found in Finley 1981, 97–115.

2. E.g., Ste. Croix 1981, 141–42, 284, 505–506; Osborne 2010; Foxhall 2002; Ismard 2017.

3. E.g., Ober 1989, 24–27; 2008; Hansen 1991, 318; cf. Meier 1990, 146.

4. Support for one or more of the preceding claims: e.g., Ste. Croix 1954/55; Meiggs 1972, 404–12; Will 1972, 171–73; Schuller 1974, 3; Galpin 1983. Critique of such claims: Finley 1981, 41–61.

5. E.g., Keuls 1993; Blundell 1995; Pomeroy 1995. Thoughtful critique of this conventional wisdom: Patterson 1998, 5–43.

6. E.g., Foxhall 1989, 1996; Katz 1992; Hunter 1994; Patterson 1994, 1998; Cohen 2000.

7. E.g., Aeschines 1.183; [Demosthenes] 59.85. Case for a distinctive female "citizenship" in Athens: Patterson 1986, 1990, 1994, 1998, esp. 107–37.

8. Foxhall 2002.

9. E.g., Mossé 1973, 114; Lintott 1982, 34; Ober 1989.

10. Quotes: Ober 1996, 4 and 18–31. Full statement of the argument: Ober 1989; cf. 2008.

11. E. Cohen 2000, 191.

CHAPTER 3

1. Self-evident state-society divide in Athens: e.g., Hansen 1998, esp. 86–91, 135–37; 2002. Public and private realms in the *polis*: e.g., D. Cohen 1991, 70–97.

2. Production of a characteristic "state effect" in modernity: Mitchell 1999.

3. E.g., Burford 1993, 15–55.

4. E.g., Aristotle, *Constitution of the Athenians* 56; Demosthenes 43.75; Isaeus 7.30.

5. E.g., Barker 1951, 5; Osborne 1985, 7–8; Morris 1987, 5; Meier 1990, esp. 20–22; Ober 1993, 129. Athens as an entirely "stateless" community: Berent 1996, 2004, cf. Cartledge 1998, 468.

6. Constant 1988, 311, 317.

7. Constant 1988, 311.

8. Finley 1985, 154–55.

9. Constant 1988, 312.

10. Constant 1988, 311, 316, 317.

11. E.g., Hansen 1996, 99; cf. Liddel 2007.

12. E.g., Burke 1992; E. Cohen 1993; Nafissi 2004, 2005; Amemiya 2007; Ober 2008, 2015.

13. The idea of a pre-modern "embedded economy" was first formulated by Karl Polanyi (1944). Seminal "substantivist" account of "the ancient economy:" Finley 1973.

14. Thucydides 2.35–46.

15. Cf. Foxhall 2007. Attempts to frame a kind of compromise between the formalist and substantivist positions: e.g., Morris 1994; von Reden 1995; Christesen 2003; Foxhall 2007; Bresson 2015.

16. Absence of a secular realm from Greek experience: Connor 1988; Bruit Zaidman and Schmitt-Pantel 1992, esp. 92–101; Samons 2000, 325–29; Blok 2009, 2011. Insistence on the reality of a secular realm in Greek antiquity: e.g., Scullion 2005.

17. E.g., Bremmer 1994, 1; Parker 1996, 1–2; Hedrick 2007, 285–86.

18. Another critique of the claim that religion was "embedded" in non-modern lifeworlds: Nongbri 2008.

CHAPTER 4

1. See especially Chakrabarty 2000.
2. Morris 1996, 2000. Quotation from 1996, 41.
3. The only sources for this event are: Aristophanes, *Lysistrata* 273–82; Herodotus 5.66–73.1; Aristotle, *Constitution of the Athenians* 20–22.1. None see it explicitly as a bottom-up "revolutionary" event.
4. Ober 1996, 32–52; 1998b, 2007.
5. Raaflaub 1998a, 1998b, 2007. The only classical source for the reforms of Ephialtes is a brief and notoriously problematic passage in the Aristotelian *Constitution of the Athenians* (25.1–2).
6. Cf. Thucydides 2.15.2; Aristotle, *Constitution of the Athenians* fr. 4, 41.2.
7. Anderson 2007.
8. The "Tyrannicides" and their significance in Athenian memory: Anderson 2003, 197–211; cf. Anderson 2007, 119–24.
9. E.g., Aristotle, *Constitution of the Athenians* 42–69.
10. Aristotle, *Constitution of the Athenians* 43.1, 61.
11. Raaflaub 1996, 158.
12. Attempt to reconstruct an ancient Greek "democratic theory:" Farrar 1989.
13. Cf. Ostwald (1996, 51): "We shall look in vain for any Greek text before or after Aristotle for [any] recognition of individual rights." Absence of liberalism from Greek thought: Cartledge 2009, 131.
14. Raaflaub 1996, 158.

1. By "mainstream" I refer to the varieties of cultural history which take their cues from Geertzian cultural anthropology, Annaliste histories of *mentalités*, German *Alltagsgeschichte*, or Italian *microstoria*. Influential examples would include: Ginzburg 1980; Le Goff 1980; Davis 1983; Darnton 1984. The genesis, gains, and limitations of history's "cultural turn:" e.g., Sewell 2005, 22–80; Spiegel 2005.
2. Cf. Weinstein's (2005, esp. 72–78) claim that cultural history fails to challenge "the historian's 'common sense.'"
3. E.g., Chakrabarty 2000, 106.
4. Holbraad 2010, 184.
5. Attempts to formulate something like a manifesto for the practice of "discursive history:" e.g., Scott 1991; Ermarth 2001.
6. E.g., Toews 1987; Palmer 1990; Jenkins 1997, 242–73, 277–312, 315–83.
7. Barad 2007, 189–222.

8. Cf. Eley 1996.

9. E.g., Chartier 1997; Biernacki 2000; Eley and Nield 2007; Bennett and Joyce 2010; cf. Coole and Frost 2010.

10. See e.g., Bynum 1999, who notes the distinctly non-modern constructions of the materiality of the human body in medieval Europe.

11. Descola 2013, 81.

12. Descola 2013, 81.

13. E.g., Harman 2002, 2010; Brassier 2007; Meillassoux 2008.

14. E.g., Latour 1993, 2005, 2013; Lynch 2012; Pickering 2016.

15. For the record, I only became aware of these calls for an ontological turn in anthropology at a relatively advanced stage of my own project, which took shape quite independently.

16. E.g., Hallowell 1960.

17. Viveiros de Castro 2003, 18. Original formulation of this alternative approach: Viveiros de Castro 1998. Fully elaborated exposition of a "cannibal metaphysics:" Viveiros de Castro 2014.

18. Holbraad 2013, 469–70.

19. Henare, Holbraad, and Wastell 2007, 9.

20. Convenient points of entry to Holbraad's project: Henare, Holbraad, and Wastell 2007, 1–31; Holbraad 2007. Fully elaborated case for an ontological turn: Holbraad 2012.

21. Descola 2013, 129–43.

22. Descola 2013, 144–71.

23. Descola 2013, 201–31.

24. Descola 2013, 172–200.

25. E.g., Halbmayer 2012a, 2012b; Scott 2013, 2014; Salmond 2014; Kohn 2015.

26. Heywood 2012, 146.

27. E.g., Deleuze and Guattari 1987. Engagement with Deleuzian thought is most extensively pursued in Viveiros de Castro 2014.

28. Viveiros de Castro 2015, 9.

29. E.g., Graeber 2015.

CHAPTER 6

1. Viveiros de Castro 1998, 470.

2. Viveiros de Castro 1998, 471.

3. Viveiros de Castro 1998, 471–72.

4. Viveiros de Castro 1998, 477.

5. Viveiros de Castro 1998, 477–78.

6. Viveiros de Castro 1998, 478.

7. Lévi-Strauss 1973, 384.

8. Viveiros de Castro 1998, 479.

9. Sahlins 1985a, 36; 1985b, 207, 214.

10. Geertz 1980, 128.

11. Geertz 1980, 129.

12. Geertz 1980, 129.

13. Marriott 1976, 1990. Premodern and non-western modes of personhood: Fowler 2004.

14. Marriott 1976, 109–10.

15. Marriott 1976, 109–10.

16. López Austin and López Luján 2001, 239–43.

17. López Austin and López Luján 2001, 242–43.

18. López Austin and López Luján 2001, 239–48.

19. Clendinnen 1991, 209.

20. Strathern 1988, 13.

21. Strathern 1988, 343.

22. Strathern 1988, 13.

23. Lienhardt 1985, 148.

24. E.g., Bird-David 1990, 1998.

25. Descola 2013, 3–27.

26. Descola 2013, 22.

27. Descola 2013, 23.

28. Jiang 2011, 58–67.

29. References: Jiang 2011, 59.

30. Wang 2000, 2. Cf. Descola's (2013, 201–31) "analogist" type of ontology.

31. Jiang 2011, 145, 174, with references to contemporary sources.

32. Jiang 2011, 142–46.

33. Jiang 2011, 61–62.

34. Jiang 2011, 157.

35. Lau 1979, 18.

36. W. Chan 1963, 20.

37. J. Chan 2005, 64.

38. Rosemont, Jr. 2005, 51.

39. Fingarette 1983, 332, 340.

40. Translation: Lewis 1954, 225.

41. E.g., Gierke 1927, 22–30.

42. Gierke 1927, 22.

43. Maitland 1901, 132–33.

44. *Commentary on the Ethics of Aristotle*, 1.1.4, from Lerner and Mahdi 1963, 274–75.

45. See especially Kantorowicz 1957, 193–272.

46. Chrimes 1949, 31.

47. Plowden 1816, 212a. Cf. Kantorowicz 1957.

48. Cited by Kantorowicz 1957, 223.

CHAPTER 7

1. E.g., Descola 2013, 32–56.
2. Descola 2013, 63–66.
3. Descola 2013, 66–68.
4. Descola 2013, 68–72.
5. Descola 2013, 68.
6. Descola 2013, 72, citing Tylor 1871, 1.
7. Descola 2013, 72–75.
8. Descola 2013, 75–78.
9. Descola 2013, 77–78.
10. Descola 2013, 73.
11. Descola 2013, 88.
12. Critical discussions of scientific materialism: e.g., Davies and Gribbin 1992; Barad 2007; Nagel 2012; cf. pp. 106–110 below.
13. The social production of nature: e.g., Haraway 1991; Latour 2004; Morton 2007.
14. Nature-less non-modern lifeworlds: e.g., Descola 1994; 2013, 32–56; Kohn 2013.
15. "Religion" as product of a modern, secular world: e.g., Asad 1993, 2003.
16. E.g., Locke 1980, 52–65; Kant 1983, 31–32. Arguments for the prevalence of individualism in non-modern worlds tend to conflate behavioral individualism with ontological individualism: e.g., Macfarlane 1978.
17. The "generation" of the "great Leviathan:" Hobbes 1996, 114 (*Leviathan* 13.17).
18. E.g., Runciman 2000.
19. Locke 1980, 9.
20. Hume 2005, 378–81, 415.
21. Paine 2000, 165–66.
22. Smith 1976, 106–107.
23. Locke 1980, 15.
24. Locke 1980, 18–30.
25. Locke 1980, 29, 66.
26. Smith 1986, 242.
27. Paine 2000, 3.
28. Spencer 1981, 384.
29. Spencer 1981, 383–434.
30. Paine 2000, 165.
31. Paine 2000, 166.
32. Hamilton et al. 1961, 38; cf. 103–104.
33. E.g., Smith 1976, 121–23, 140–43; 1986, 265, 323.
34. E.g., Mirowski 1988.
35. Mitchell 1998, 2008.

36. The protection of property rights as the primary purpose of government: e.g., Hamilton et al. 1961, 41–60, 77–99; Mill 1967, 4–5; Locke 1980, 29; Smith 1986, 293–94; Hume 2005, 373–416.
37. Liberal market freedom and its formation: e.g., Harcourt 2011.
38. Descola 2013, 81.

CHAPTER 8

1. Malin 2001, 15.
2. Barad 2007, 106.
3. On the quantum revolution in physics, I have found the following especially helpful: Heisenberg 1958; Bohr 1961; Malin 2001; Barad 2007; Rosenblum and Kuttner 2011.
4. Bohr 1961, 39–40.
5. Bohr 1961, 72. An accessible introduction to Bohr's "philosophy-physics:" Barad 2007, 97–131.
6. Barad 2007, 17.
7. Quotation originally reported in Petersen 1963, 12.
8. Barad 2007, 110.
9. Aside from Barad (2007), the only other authority I know of who has made a case for something like a quantum human science is Alexander Wendt (2015).
10. Marx 1973, 84. Though not published until 1939, the notes which comprise the *Grundrisse* ("Foundations [of the Critique of Political Economy]") were originally written during the years 1857–1861.
11. Seminal contributions to this larger project: Foucault 1970, 1972.
12. Liberal governmentality and biopolitics: e.g., Foucault 1990, 135–59; 2008, 1–73; Burchell et al. 1991; Rose 1999.
13. E.g., Mitchell 1988, 1999, 2008; Jessop 2008; Brown 2010
14. E.g., Latour 1993, 2005.
15. E.g., Bordo 1987, 1993, 1999; Butler 1990, 1993.
16. E.g., Haraway 1991, 2003, 2008; Oyama 2000; Keller 2010; Kirby 2011.
17. E.g., Pickering 1981, 1984, 1995; Bachelard 1984; Latour and Woolgar 1986; Latour 1987; Canguilhem 2008.
18. E.g., Winch 1958, 1964; Kuhn 1962; Hesse 1966, 1974.
19. Poovey 1998.
20. Daston and Galison 2007; cf. Daston 2000.
21. E.g., Rorty 1979, 1989; Feyerabend 1975, 1987; Vattimo 2011.
22. Barad 2007, 33. Since an "interaction" can only take place between previously distinct agencies, Barad coins the term "intra-action" to signify "the mutual

constitution" of agencies that do not already exist as individual elements, that "are only distinct in relation to their mutual entanglement."

23. Barad 2007, 192.

24. Barad 1998, 2007. Quotations: Barad 2007, 33.

25. E.g., Strathern 1992a, 1992b; Graeber 2001, 2011; Descola 2013; Kohn 2013; Rosemont, Jr. 2015.

26. E.g., Spivak 1988; Asad 1993, 2003; Prakash 1994; Chakrabarty 2000, 2002; Bhambra 2007, 2011.

27. The rather ungainly locution "meta-metaphysics" is in fact already current in contemporary philosophy: e.g., Chalmers et al. 2009.

CHAPTER 9

1. This possibility is explored further in the Conclusion.

CHAPTER 10

1. To my knowledge, the following account represents the first explicit attempt by a modern historian to document the metaphysical underpinnings of the Athenian *politeia*.

2. Aristotle, *On the Soul* 411a7.

3. Cole 2004, 14–15.

4. Gould 1985, 7.

5. Cf. Cole 2004, 37–38: "The gods belonged to the natural world and were therefore necessarily considered prior to the *polis*, even when new cities were founded in new territories."

6. Cf. Parker 2005, 2.

7. E.g., Sophocles, *Oedipus at Colonus* 1006–1007; St. Paul, *Acts* 17:22.

8. Overview of "unlicensed religion" and "magic" in Athens: Parker 2005, 116–35.

9. The principal gods of the Athenians and their respective life-sustaining functions: e.g., Parker 2005, 387–451; Mikalson 1983, 18–26; cf. Chapter 11 below.

10. Text: *Inscriptiones Graecae* II² 4960. Discussion of the monument, along with the whole process of "introducing" the "new god" Asclepius to Attica: Garland 1992, 116–35.

11. Others, like Parker (2005, 389–90), retain this modern dualist distinction.

12. Visible "effects" of divine activities in Greek experience: Parker 2011, 2–13. Divine ephiphanies: e.g., Platt 2011.

13. Divine tableau on the Parthenon frieze: Neils 1999.

14. Athens/Attica as *theophiles*: e.g., Isocrates 12.125; Plato, *Menexenus* 237c. More gods in Athens: Garland 1992, 14; Parker 2005, 397; cf. Xenophon, *Constitution of the Athenians* 3.2; Pausanias 1.17.1, 1.24.3.

15. Cf. Demosthenes 4.35–36.

16. Gods who governed political practices: Parker 2005, 403–408.

17. Samons 2000, 327. Cf. also, e.g., Mikalson 1983, 13–17.

18. Samons 2000, 326.

19. E.g., Finley 1983, 94; Lewis 1990, 259; Dowden 2007, 41. Studies more willing to consider Greek cosmologies on their own ontological terms include: e.g., Gould 1985; Garland 1992; Cole 2004; Parker 2005; Larson 2007; Blok 2011, 2014.

20. Most influential case for a "system" of "*polis* religion:" Sourvinou-Inwood 1990.

21. E.g., Asad 1993, 2003; cf. Nongbri 2013.

22. Cf. Barton and Boyarin 2016.

23. Demosthenes 23.40; cf. *Inscriptiones Graecae*, I^3 104.

24. The *polis* defined as the totality of its *hiera kai hosia*: e.g., Demosthenes 59.104; Isaeus 6.47; Isocrates 7.66; Lycurgus 1.77; Lysias 30.25; Plato, *Republic* 344a; Thucydides 2.52.3–4; Xenophon, *Ways and Means* 5.1–4.

25. The logic of ritual purification: e.g., Burkert 1985, 269; Cole 2004, 35–36.

26. Burkert 1985, 271.

27. E.g., Blok 2014; cf. 2009, 2011; Peels 2014.

28. E.g., Aeschylus, *Prometheus Bound* 527–29; Sophocles, *Oedipus at Colonus* 469–70.

29. E.g., Blok 2010.

30. Samons 2000, esp. 30–50.

31. E.g., Aristophanes, *Lysistrata* 743–44.

32. Demosthenes 60.4. Other sources for Athenian autochthony include: Aristophanes, *Wasps* 1076; Euripides, *Ion* 589–90; Herodotus 1.56.2, 7.161.3; Hyperides 6.7; Isocrates 4.24, 8.49, 12.124–5; Lysias 2.17; Plato, *Menexenus* 237b; Thucydides 1.2.5, 2.36.1. To my knowledge, stories of the autochthonous origins of the Athenians were never seriously challenged by other Greeks.

33. Lysias 2.17.

34. Cecrops: e.g., Aristophanes, *Wasps* 438; Euripides, *Ion* 1163–64; cf. Gantz 1993, 236. Erechtheus: e.g., Herodotus 8.55; Homer, *Iliad* 2.547–8; Dionysius of Halicarnassus, *Roman Antiquities* 14.2; Erichthonius: e.g., Euripides, *Ion* 20–21. Vase images: e.g., Loraux 1993, 29–31.

35. Athenian people as *Erekhtheidai*: e.g., Homer, *Iliad* 2.547; Pindar, *Isthmian* 2.19; Sophocles, *Ajax* 202. Cf. Athenians as descendants of Cecrops (*Kekropidai*): e.g., Herodotus 8.44.2.

36. Isocrates 4.24–25.

37. E.g., Rosivach 1987; E. Cohen 2000, 79–103.

38. On the paternal and maternal properties of land, e.g., L'Homme-Wéry 2000; Cole 2004, 1–2.

39. Plato, *Menexenus* 237d–238a. Similar claims are made in e.g., Demosthenes 60.4–5. While the passage in the *Menexenus* is part of a speech which may well parody the orations delivered at the annual funerals for the war dead, any humor here would obviously depend on the use of ideas and sentiments that were in fact commonplace

in the genre. Interpretations of the tone and purpose of this unusual work vary: e.g., Salkever 1993; E. Cohen 2000, 100–102.

40. Plato, *Menexenus* 238a–b.

41. Isocrates 12.125: *ten trophon ex hes per ephusan . . . stergontas auten homoios hosper oi beltistoi tous pateras kai tas meteras tas auton.*

42. E.g., Parker 2005, 426–28.

43. Notorious violation of this injunction after Arginusae in 406: Xenophon, *Hellenica* 1.6.35–7.35.

44. Plato, *Menexenus* 237b–c.

45. E.g., Plutarch, *Solon* 12, on the expulsion of the bones of the "cursed" Alcmeonid family from Attica (late seventh century?). Cf. Cole 2004, 35–36; Sophocles, *Oedipus Tyrannus* 14–57; *Antigone* 998–1022.

46. Formal "trials" of such objects in Athens: e.g., Aristotle, *Constitution of the Athenians* 57.4.

47. Xenophon, *Hellenica* 1.7.22.

48. Lycurgus 1.1.

49. "Sacred" and "public" land in Attica: Papazarkadas 2011.

50. "Ritual space" and its various sub-categories: Cole 2004, 30–65.

51. Acropolis: e.g., Thucydides 2.17; Cole 2004, 58–59. Grove of the *Semnai Theai*: Sophocles, *Oedipus at Colonus* 37, 39, 126. The *hiera orgas*: Papazarkadas 2011, 244–59. The *moriai*: Papazarkadas 2011, 260–84.

52. Estimates of the proportion of *hiera* land in Attica: e.g., Andreyev 1974, 43 (ca. 10%); Papazarkadas 2011, 96–97 (ca. 4%).

53. Papazarkadas (2011, 212–36) concludes that *demosia* lands were mostly uncultivated "zones in mainly marginal areas . . . such as those located along circuit walls."

54. E.g., Aristotle, *Politics* 1274b71; Thucydides 7.77.7. For further references, see Hansen 1998, 56–57.

55. A *polis* as a human body: e.g., Aristotle, *Politics* 1302b34–1303a2.

56. *Demos* as a unitary corporate person: Anderson 2009. Cf. Thucydides 3.82.2, where a *polis* is said to have its own "mind" (*gnome*). Demos was typically figured as a single, mature, male figure: e.g., Aristophanes, *Knights*; Pausanias 1.3.3; Pliny, *Natural History* 35.69; *Supplementum Epigraphicum Graecum* 12.87; cf. *Inscriptiones Graecae* II² 844.38–39. Catalogue of personified Demos images: Smith 2003. Cult of Demos and the Nymphs (early classical era?): *Inscriptiones Graecae* I³ 1065.

57. Aristotle, *Politics* 1253a19–29.

58. Ostwald 1996, 56.

59. Hence too there was no routine "naturalization" procedure in Athens. Demos would occasionally incorporate non-Athenians to honor their extraordinary contribution to the life of the social body. E. Cohen (2000, 49–78) argues that children of established non-Athenian "residents" (*astoi*) in Attica were commonly accepted as Athenians.

60. Symbiosis of *polis* and *oikos*: e.g., Todd 1993, 201–31; Hunter 1994.
61. Thus, a sentence of exile in Athens and elsewhere in the Greek world meant the excision of an *oikos* from the body of *polis* and thus its effective annihilation.
62. The *polites/idiotes* distinction: e.g., Rubinstein 1998.
63. Cf. Gill 1996 on relational Greek selfhood.
64. Conventional belief that some form of *ephebeia* ("cadetship") prevailed in Athens since at least the early classical era, with the practice becoming more formalized after the decree of Epicrates in 335 BC: e.g., Steinbock 2013, 294–95.
65. "Cecropian agora:" Plutarch, *Life of Cimon* 4.6. Prytaneion sources: Thompson and Wycherley 1972, 166–74. Location: Dontas 1983. Structure: Schmalz 2006.
66. E.g., Demosthenes 19.303.
67. E.g., Pausanias 1.18.2; Philochorus, *Fragments of the Greek Historians* 328 F105.
68. E.g., Merkelbach 1972; Hadjisteliou-Price 1978, 105, 113; Kearns 1989, 23–27, 57–63; Parker 2005, 433–35, with nn.63–65. Cecropid sisters named "after the things that make crops grow": Stephanus of Byzantium, s.v. *Agraule*. Aglauros and Pandrosos served by the same priestess: Parker 2005, 216.
69. The weapons in question were most likely a shield and a spear (cf. Aristotle, *Constitution of the Athenians* 42.4), previously stored on the consecrated acropolis (cf. Plutarch, *Moralia* 852C).
70. The text of the oath is reconstructed from Lycurgus 1.76 and an inscription from Acharnae, north of Athens (e.g., Rhodes and Osborne 2003, 440–49). Cf. Siewert 1977.
71. *Politeia* as an all-inclusive "way of life": e.g., Aristotle, *Politics*; Plato, *Republic*; Thucydides 2.34–47. The *politeia* of a *polis* is thus its "life" or "soul:" e.g., Aristotle, *Politics* 1295a40–b1; Isocrates 7.14.

CHAPTER 11

1. E.g., Foucault 1990, 135–59; 2008, 1–73; Burchell et al. 1991; Rose 1999.
2. The "work" of the gods in Attica: Parker 2005, 387–451.
3. Mountain-top shrines of Zeus the weather god in Attica included those of Zeus Ombrios ("Rain-Bringer") and Zeus Proopsios ("Look-Out") on Mt. Hymettos and those of Zeus Ombrios/Apemios ("the Unharmful One") and Zeus Semaleos ("Sign-Giver") on Mt. Parnes: Pausanias 1.32.3; Langdon 1976, esp. 78–112; Parker 1996, 29–33. Cf. Zeus and ploughing: Parker 2005, 417n.4.
4. Main shrine of Ge (Olympia) in Athens, south of the acropolis: Thucydides 2.15.4; cf. Pausanias 1.18.7; Plutarch, *Theseus* 27.6.
5. Poseidon associated with springs and, as Poseidon Phytalmios, the growth of plants: e.g., Aeschylus, *Seven Against Thebes* 308–10; *Inscriptiones Graecae*, II² 5051; Plutarch, *Convivium Septem Sapientium* 15.158d. Shrine of the eponymous river god Cephisus, where the altar was apparently shared with a other life-sustaining deities and nymphs, like Hestia, Apollo Pythios, Eileithyia, and Kallirhoe: Parker 2005,

430, with nn.48–49. Cf. "life-giving" (*biodoroi*) nymphs, daughters of river god Inachos in the Argolid: Aeschylus, fr. 168.17. Sun and Seasons ritually implicated in Athenian Skira and Thargelia festivals: Parke 1977, 157; Parker 2005, 203. River gods, nymphs, and other "nature deities:" e.g., Larson 2007.

6. Cult of Demeter at Eleusis, her primary home in Attica: e.g., Clinton 1992. The Thesmophoria, the principal "fertility" festival of Demeter: e.g., Parke 1977, 82–88. Grapevine/wine-related festivals of Dionysus in Athens, like the Oschophoria, Lenaia, and Anthesteria: Parke 1977, 77–80, 104–106, 107–24. On Athena, the sacred olive, and its descendant trees, e.g., Herodotus 8.55; Philochorus, *Fragments of the Greek Historians* 328 F67; Lysias 7; Papazarkadas 2011, 260–84. Apollo venerated as nurturer of vegetal life at the Thargelia and Pyanopsia festivals: Parke 1977, 74–77, 146–49; Parker 2005, 203–206.

7. E.g., Homer, *Iliad* 14.490–91; *Homeric Hymn to Hermes* 491–95.

8. Iconographic associations of these and other gods, like Dionysus and Heracles, with the cornucopia, the horn of plenty: Parker 2005, 418–23.

9. Early king Erichthonius mandated preliminary sacrifices to Ge Kourotrophos at Athenian festivals: Suda s.v. *kourotrophos ge, k* 2193. Evidence for these offerings in sacrificial calendars of Erchia, Thorikos, Marathon, and the Salaminioi *genos*: Parker 2005, 426–28, with table 2. Sanctuary of Ge Kourotrophos in Athens: Pausanias 1.22.3.

10. A *genetyllis* was a "spirit responsible for procreation," associated with Artemis, Aphrodite, and erotic behavior: Suda s.v. *genetyllis, g* 141. Genetyllides worshipped at the sanctuary of Aphrodite Kolias on the western coast of Attica: Pausanias 1.1.5. Spring sacred to Aphrodite on Mt. Hymettos, where "women who drink give birth easily, and those who are infertile become fertile": Suda s.v. *kullou peran, k* 2672. Vulva and breast votive plaques at Aphrodite sanctuaries represent requests for help with childbirth and lactation, not for healing of those body parts: Forsen 1996, 133–34. Kalligeneia and absence of Greek categorical distinction between human and non-human forms of fertility: Parke 1977, 87–88. Eileithyia responsible for sustaining children through the notoriously precarious first five or six years of life: Parker 2005, 428. Eileithyia's main shrine in Athens: Pausanias 1.18.5.

11. Votives from her Brauron sanctuary suggest that Artemis cared for both boys and girls, presumably as part of her larger jurisdiction over the young of all species: Parker 2005, 428. Aglauros was protector of male youths, especially ephebes (see previous chapter). As protector of young females, Pandrosos shared a priestess with Ge Kourotrophos and was guardian of the *arrhephoroi*. Cf. Pausanias 1.27.3; Parke 1977, 140–43. *Kharis* and the *Kharites* ("Graces"): MacLachlan 1993.

12. Parker 2005, 440–43.

13. "Household religion:" e.g., Boedeker 2008; Faraone 2008. Zeus Ktesios: e.g., Aeschylus, *Suppliants* 443–44; Autokleides, *Fragments of the Greek Historians* 353

F1; Isaeus 8.16; Parker 2005, 15–16. Zeus Herkeios: e.g., Aristotle, *Constitution of the Athenians* 55.3; Herodotus 6.68; Sophocles, *Antigone* 487; Parker 2005, 16–18.

14. Center of the *oikos*: e.g., Plato, *Laws* 740b, 771c. Introduction of brides to Hestia: e.g., Theopompus, *Fragments of the Greek Historians* 115 F15. *Amphidromia* ritual, which introduced the new-born to the household: Hamilton 1984. Presentation of new slaves: e.g., Aristophanes, *Wealth* 768–69, 788–99; Demosthenes 45.74.

15. Statues of Hermes and pillars of Apollo Aiguieus and Hekate: Parker 2005, 18–19.

16. Introduction of Asclepius to Athens: e.g., Garland 1992, 116–35. Amphiaraus: Kearns 1989, 147. Athena Hygieia on the acropolis: e.g., *Inscriptiones Graecae*, I³ 506. Athena ladles "wealth-health" (*plouthugieia*) over the *polis* as a whole: Aristophanes, *Knights* 1091. Votives of breasts and female genitalia dedicated to Aphrodite and Artemis Kalliste: Parker 2005, 412, with n.99. Athenian shrine of Apollo Paion (protected *polis* as a whole from plagues, etc.) and Athena on the island of Delos: *Inscriptiones Graecae*, I³ 1468 *bis*.

17. Parker 2005, 409, with nn.89–91. Athena Ergane and veneration at Chalkeia festival: Sophocles, fr. 760 Nauck; Parke 1977, 92–93.

18. Parker 2005, 409–11.

19. Hermes in the agora: Wycherley 1957, 102–108.

20. These and other "political gods:" Parker 2005, 403–408.

21. Parker 2005, 397–403.

22. Apaturia: Parke 1977, 88–92. Thargelia: Parke 1977, 146–49.

23. References: Wycherley 1957, 21–31. Stoa: Travlos 1971, 527–33.

24. Metroon in the agora: Travlos 1971, 352–56.

25. Rosenzweig 2004, 13–28; Pausanias 1.22.3.

26. Zeus Polieus honored at Dipoleia festival: Parke 1977, 162–67; Hurwit 1999, 190–92. Cult of Poseidon-Erechtheus, whose altar was located inside the Erechtheion, the principal temple of Athena Polias: Hurwit 1999, 32–33, 38, 39, 56, 200–202.

27. Quotations: Euripides, *Children of Heracles* 770-2; Solon, fr. 4.1-4 West. Athena as supreme patron-protector of the *polis*: Herington 1955, 55–58.

28. Hence, for example, in decrees which honored their services, Athenian priests and priestesses were routinely praised for helping to secure this "health and safety" of the *polis* through their ritual performances: Parker 2005, 95–96; Lambert 2012, 74–75.

29. Reconstruction of text (here slightly adapted): Rhodes 1972, 37.

30. Aristotle, *Constitution of the Athenians* 43.6.

31. Laws relating to ritual performance and *ta hiera*: Todd 1993, 307–15.

32. Priests and others who performed sacrifices provided reports of the good omens their rites had secured to support their claims to honors from the community: Parker 2005, 95–96.

33. Wide range of Athenians authorized to make offerings for the "health and safety" of the *polis* as a whole: Parker 2005, 89–115. Slaves and metics participated in

festivals like the Kronia, Anthesteria, and Panathenaia: e.g., Parke 1977, 30, 117; Maurizio 1998.

34. E.g., Blundell and Williamson 1998; Dillon 2002; Brulé 2003, 6–42; Goff 2004; Stehle 2012.

35. E.g., Lewis 2002, 43: "For those who find ancient attitudes towards women depressing, the field of ritual is always a saving grace." Cf. Brulé 2003, 5: "[E]veryone says that [religion] is what matters most [in ancient experience] and then promptly forgets this priority."

36. The following discussion focuses mostly upon the ritual contributions of females to the life of the *polis* as a whole. Females and "household religion:" e.g., Faraone 2008; cf. pp. 166–67 below.

37. Euripides, fr. 494.9–22.

38. Aristophanes, *Lysistrata* 638–51.

39. Arrhephoria: e.g., Pausanias 1.27.3; Parke 1977, 140–43. The *archon basileus* was "the individual who had the highest responsibility in religious affairs:" Parker 2005, 91.

40. Parke 1977, 140–41.

41. Aristotle, *Constitution of the Athenians* 54.7. Brauron sanctuary, Brauronia, and *arkteia*: e.g., Deubner 1932, 207–208; Parke 1977, 139–40; Kahil 1983; Osborne 1985, 154–76.

42. E.g., MacLachlan 2012, 117–18. *Kanephoros* in Athenian and Greek art: Roccos 1995.

43. Appointment of priests in Athens: e.g., Aleshire 1994. "Social construction" of priests and priestesses in Athens: Lambert 2012.

44. Plato, *Statesman* 290c-d.

45. Significant exceptions included the major panhellenic cults of Zeus at Dodona and Apollo Pythios at Delphi, both of which were served by priestesses, as noted above in the passage from *Melanippe Captive*.

46. E.g., Connelly 2007, esp. 59–69 (priestesses of Athena Polias and Demeter and Kore). As Connelly (2007, 64) notes: "So central were priestesses to Athenian society that their names were household words and they were fair game for jokes and for portrayal in the theater."

47. Adonia: e.g., Dillon 2002, 162–69. Skira (for Demeter and Kore): Parke 1977, 156–62. Stenia (for Demeter and Kore): Parke 1977, 88, 156, 188; Dillon 2002, 109.

48. Stehle 2007; 2012, 195. Other discussions, some of them quite explicitly "translating" the meaning of the festival into modernist, secularist terms, include: Parke 1977, 82–88, 158–60; Detienne 1989; Dillon 2002, 110–20; Goff 2004, 125–38, 205–11; Parker 2005, 270–83.

49. Maurizio 1998, 302. As Maurizio (1998, 302–303) goes on to note, the east end of the Parthenon Frieze, showing the front of the Panathenaic procession and (probably) the presentation of the *peplos*, is dominated by female figures.

50. Maurizio 1998, 303–304.
51. Maurizio 1998, 297–98.

CHAPTER 12

1. Aristotle, *Politics* 1252b12, 1253b1–3; cf. *Economics* 1343a; Plato, *Laws* 720e–721e.
2. Foxhall 1989, 22.
3. Foxhall 1989, 43.
4. E.g., D. Cohen 1991, 1995; Burford 1993; Todd 1993, esp. 201–31; Hunter 1994; Patterson 1998; E. Cohen 2000.
5. Davidson 2011, 183.
6. Xenophon, *Oeconomicus* 7.18–19.
7. Xenophon, *Oeconomicus* 7.20–27.
8. Xenophon, *Oeconomicus* 7.30.
9. Xenophon, *Oeconomicus* 7.35–37.
10. Aristotle, *Politics* 1254b15.
11. Aristotle, *Economics* 1343b.
12. Aristotle, *Economics* 1343b–44a.
13. Lysias 1.6–7.
14. Hunter 1994, 36. Survey of the evidence: Hunter 1994, 33–37.
15. Survey of the evidence: Brock 1994.
16. Brock 1994, 346.
17. Hunter 1994, esp. 9–42.
18. Foxhall 1989.
19. Foxhall 1996.
20. E. Cohen 2000, esp. 30–48.
21. e.g., Patterson 1981, 1994, 1998.
22. Almost forty years ago, the feminist-socialist historian Sheila Rowbotham ([1979] 2006) was already seriously questioning the general use of patriarchy as a historical category, critiquing its reductive analytical bluntness, its dubious universalism, its biological essentialism, and its sweeping denial of agency to females across time and space.
23. Morris and Powell 2010, 29.
24. Pomeroy et al. 2008, 253.
25. Blundell 1995, 11.
26. James and Dillon 2012.
27. Henry and James 2012, 85.
28. Levick 2012, 98, 105.
29. Hesiod, *Works and Days* 53–105; *Theogony* 590–602; Semonides fr. 7 West. According to Martin West (1978, 165–66), earlier, more definitive stories about Pandora represented her simply as the wife of Prometheus and bestower upon humankind of "all gifts" from the gods, much as her name suggested.

CHAPTER 13

1. Perhaps the nearest thing to mechanisms of "universal education" in Athens would be the tragedies and comedies produced at festivals of Dionysus. Educational functions of Athenian drama: e.g., Gregory 1991.

2. Phratries: Lambert 1998. *Gene*: Bourriot 1976; Roussel 1976; Parker 1996, 284–327; Lambert 1999. Other corporations: Parker 1996, 333–42.

3. Parker: 1996, 328–32.

4. Mikalson 1977; Whitehead 1986, 176–222.

5. Evidence for routine female participation in deme rituals: Jones 1999, 123–33. Evidence for the term *demotis* ("female deme member"): Aristophanes, *Lysistrata* 333.

6. Full text: Daux 1983.

7. E.g., sacrifice to Poseidon at Sounion: Whitehead 1986, 196.

8. Whitehead 1986, 198–99.

9. E.g., Osborne 1985, 178–81.

10. Evidence survives of the calendars of various demes: e.g., Erchia; Teithras; Marathon; Eleusis. Evidence for "agrarian Dionysia" festivals and eponymous hero honors in demes: e.g., Dow 1968; Mikalson 1977; Whitehead 1986, 185–222.

11. Summary discussion of marriage in Athens: e.g., Cox 2011.

12. Patterson 1998, 108–109.

13. Marriage thus simply a "given" in Athenian laws: Patterson 1998, 112. Legitimacy and health of marriages as matters of common knowledge among neighbors: e.g., Isaeus 3.13–15, 80, 6.10–11, 8.18–20.

14. Correlations between marriage and locality (based on attested demotics): Cox 1998, 52–63, with tables 2a-d.

15. Osborne 1985, 137; Cox 1998, 3–9.

16. Osborne 1985, 135–36.

17. Debates about the nature, size, and inclusivity of phratries: e.g., Hedrick 1991; Lambert 1998; Jones 1999, 195–220.

18. E.g., Carawan 2010, 389.

19. *Inscriptiones Graecae* II^2 1237.68–113. Discussion: e.g., Hedrick 1991; Lambert 1998, 285–93; Carawan 2010.

20. *Inscriptiones Graecae* II^2 1237.88–113, with translation from Lambert and Schuddeboom 2015.

21. Aristotle, *Constitution of the Athenians* 42.1–2.

22. The council of 500 merely verified visually that candidates were of suitable age. "Pericles' citizenship law" (451), mandating that a true-born Athenian must have two Athenian parents: Patterson 1981. Small size of demes and mutual familiarity of members: Osborne 1985, 44; Whitehead 1986, 223–34. Osborne (1985, 72–73) suggests that demes may have only documented admissions in meeting minutes, keeping no formal membership lists.

23. Aeschines 1.127.

24. Hunter 1994, 116–18.

25. Catalogue of ancient textual references to gossip topics: Hunter 1994, 118–19.

26. Aristotle, *Constitution of the Athenians* 45.3, 55.2–4. Would-be council members were just scrutinized by their predecessors; would-be archons were examined by council members and by a panel of judges. Speeches delivered at courtroom scrutinies of archons: Lysias 16, 24, 25, 31.

27. E.g., Aeschines 1.153; Demosthenes 52.1, 54.38; Hyperides 1.14, 4.23.

28. It was conventional wisdom in Athens that a slave's testimony was legitimate only if it was extracted under torture, though apparently this procedure was rarely used: e.g., Todd 1993, 96, with n.22.

29. Lysias 1.23–25.

30. Even if Euphiletus's story was not entirely true, it presumably offered a version of events that would have seemed plausible to the judges, and thus, like many other such forensic passages, still provides solid evidence for Athenian norms.

31. Herodotus 9.5.2–3.

32. House-razing to extirpate a "polluted" *oikos* from a *polis*: Connor 1985. Examples: Craterus, *Fragments of the Greek Historians* 342 F5 and F17. Spontaneous "popular justice" in the Greek world: Forsdyke 2012, 144–70.

33. Laws authorizing Athenians to kill various kinds of miscreants on the spot: e.g., Andocides 1.96–98; Aristotle, *Constitution of the Athenians* 57.3 Demosthenes 23.28; Isaeus 8.44; Lysias 1.30. 6.18, 13.68.

34. Apprehension procedures: Hunter 1994, 134–39.

35. Lysias 1.37.

36. Emergency measures: e.g., Andocides 1.11–15; Xenophon, *Hellenika* 1.7.3.

37. Attested *dike* procedures: Todd 1993, 102–105. Attested *graphe* procedures: Todd 1993, 105–109. Other attested procedures which were designed to protect the *polis* as a whole from harm (e.g., *euthuna, eisangelia, apophasis*): Todd 1993, 112–16.

38. Hunter 1994, 129–43.

39. Legal and other responsibilities of these various poliadic actors: Aristotle, *Constitution of the Athenians* 52.1 (the Eleven), 55–59 (archons), 63–69 (judges).

40. Lanni 2006.

41. E.g., a fine of 1,000 drachmas for prosecutors in *graphai* who failed to secure one-fifth of the judges' votes. Other procedures to prevent abuse of the laws: e.g., *graphe adikos heirkhthenai* ("indictment for unjust arrest"), *graphe nomon me epitedeion theinai* ("indictment for improper application of the laws"), and *graphe sukophantias* ("indictment for the sale of legally compromising information" about someone).

42. Aeschines 1.2; cf. Demosthenes 21.8, 24.8, Lysias 7.20. Acceptability of personal enmity as motivation for prosecution: Hunter 1994, 127–28; Alwine 2015, 55–116.

43. Lysias 1.25–26.

44. E.g., Lysias 3, where strangers act as peace-keepers, protectors of the injured, and as trial witnesses.

45. E.g., Hunter 1989; Foxhall 1996.

46. Formation of the tribes in the late sixth century: Anderson 2003, 34–42.

47. Aristotle, *Constitution of the Athenians* 43–46.

48. Van Wees 2004, 99–101.

49. E.g., Aristotle, *Constitution of the Athenians* 42.2–3 (*ephebeia*); 63–66 (courts); Thucydides 2.34 (annual funeral for the dead); Jones 1999, 166 (*polis* boards); Stanton and Bicknell 1987 (assembly meetings).

50. Dithyrambic choruses: Wilson 2000. Panathenaic contests for tribal teams of Athenian "warriors" (*polemisterioi*): Anderson 2003, 165–69.

51. Anderson 2003, 34–42, 124–34.

52. Traill 1986, 79–92; Jones 1999, 174–82.

53. Jones 1999, 164–69.

54. Anderson 2003, 125–31. Locations of tribal shrines: Jones 1999, 156–61.

55. Delphic selection of heroes: Aristotle, *Constitution of the Athenians* 21.6.

56. E.g., "the tribe of Cecrops" could be referred to as *hoi Kekropidai* ("the descendants of Cecrops"). Evidence: Demosthenes 58.18; Cleidemus, *Fragments of the Greek Historians* 323 F22; Parker 1996, 121n.68. More commonly, the tribes were referred to using a feminine singular patronymic adjective, as e.g., *he Kekropis* (sc. *phule*), meaning "the tribe descended from Cecrops."

57. Demosthenes 60.27–31.

58. Parker 1996, 121.

CHAPTER 14

1. Berent 1996, 2004; cf. Cartledge 1998.

2. States are exclusively modern phenomena, realizable only in a modern metaphysical environment: Anderson, Forthcoming b.

3. This mainstream, broadly Weberian account of the state: e.g., Carneiro 1970; Haas 1982; Evans et al. 1985; Mann 1986; Tilly 1992; Blanton and Fargher 2008. Critical overview of various contemporary currents in state theory: Anderson 2009, 2–8.

4. E.g., Ober 1989, 1996, 2008.

5. E.g., Thucydides 2.15.2.

6. Anderson 2009.

7. Hobbes 1996, 106–10 (16.1–13).

8. This distinction between ancient and modern ontological presuppositions should have been made clearer in Anderson 2009.

9. Demosthenes 21.32.

10. Evidently, advisors briefly shed their *polites* personae when they made such proposals, then subsequently reassumed them when they voted on the issue as members of Demos.

11. E.g., Aeschines 1.173; Demosthenes 21.215–16, 59.107.

12. See e.g., Aeschines 1.173; Demosthenes 21.215–16, 59.107. Demos was *palaikhthon* ("inhabiting its land since ancient times"): Aeschines 3.190. Primordiality of Athenian *demokratia*: Anderson 2007.

13. Demos "in exile": e.g., Aeschines 2.176, 3.187; Demosthenes 19.280; Isocrates 7.16, 15.232.

14. E.g., Andocides 1.81; Lysias 10.4, 12.57–58, 13.47–48.

15. E.g., Andocides 1.96–98; Aristotle, *Constitution of the Athenians* 8.4; Demosthenes 24.144, 46.26; Hyperides 4.7–8; Lycurgus 1.125–26; *Supplementum Epigraphicum Graecum* 12.87. Analysis: Anderson, Forthcoming a.

16. E.g., Ober 2003, 224.

17. Cf. Aeschines 1.5; Lysias 13.16.

18. Lycurgus, *Against Leocrates* 1.147.

19. Lycurgus, *Against Leocrates* 1.150.

20. Aeschines, *Against Ctesiphon* 3.252.

21. E.g., Nietzsche 1976, 160; Lenin 1992. On the modern "statist imaginary:" Neocleous 2003.

22. Mann 1986, 114.

23. Council as coordinator between Demos and poliadic actors: e.g., Aristotle, *Constitution of the Athenians* 44–49; Rhodes 1972, 88–143.

24. Stone 1989, 197. Similar views among specialists: e.g., Hansen 1991, 76–78. Views of moderns surveyed, from Voltaire to Martin Luther King: Wilson 2007. Sources for indictment, trial, and execution: e.g., Plato, *Euthyphro; Apology; Crito; Phaedo.*

25. Cf. Todd 1993, 126–27.

26. Of the two favored modes of execution, hemlock poisoning was self-administered by the condemned, while the crucifixion-like "boarding" (*apotumpanismos*) caused a gradual death by asphyxiation and/or exposure.

27. Diogenes Laertius, *Lives of the Philosophers* 2.40.

28. Given that Demos was formally committed to an *amnesteia* ("forgetting") regarding the regime of the Thirty, Socrates could not have been prosecuted directly for such associations.

29. E.g., Garland 1992, 136–51; Parker 1996, 202–203; Burnyeat 1997; Cartledge 2009, 76–90.

30. Socrates' popular reputation: e.g., Plato, *Apology* 19b–c. Even so, only 280 judges apparently voted to condemn Socrates, with 220 voting for his acquittal. But when Socrates suggested that he should receive free meals for life as a benefactor of the *polis* and pay just a modest fine as an alternative to the death penalty proposed by the prosecutor, a larger majority then voted for his execution.

31. Socrates declined to follow the apparently common practice of bribing the Eleven to secure his escape: Plato's *Crito* 44b–c.

32. Death scene: *Phaedo* 115a5–118a17.

33. Later Athenians seem to have been generally untroubled by the decision to execute the philosopher: cf. Aeschines 1.173. Relative freedom of Athenians to express "dissent": Ober 1998a, 3–8.

34. Demos as ultimate authorizer of all *polis*-wide rules, even after it had delegated, say, the drafting of *nomoi* to smaller boards of judges: Hansen 1987, 178n.656.

35. Cf. decree of 410/9 BC outlining competence of Demos (probably a republication of a late sixth/early fifth century measure): *Inscriptiones Graecae* I³ 105; cf. Ostwald 1986, 31–40. Inventory of all known decrees published by Demos (403–322 BC): Hansen 1987, 108–18.

36. Rhodes 1972, 37–38; cf. Aeschines 1.23.

37. Aristotle, *Constitution of the Athenians* 43.6.

38. Parker 2005, 89–90.

39. Rhodes 1972, 127–34. A majority of the lands administered by Demos or by smaller bodies like tribes and demes were considered to belong ultimately to gods: Papazarkadas 2011, 16–62. Division of labor in ritual matters between priests and poliadic actors: Parker 2005, 89–115.

40. Aeschines 1.23.

41. Diplomacy in ancient Greece: e.g., Adcock and Mosley 1975. Grain imports: Moreno 2007. Proxenies in the Greek world: e.g., Mack 2015. Athenian *proxenoi* in the fifth century: Walbank 1978. Honors awarded to non-Athenian benefactors: e.g., Engen 2010.

42. Rhodes 1972, 113–22.

43. Aeschines 1.23.

44. Rhodes 1972, 122–27. Relatively modest quantities of "public property" in Attica: e.g., Lewis 1990; Papazarkadas 2011.

CHAPTER 15

1. Critique of the "culture" of modern economics from a comparative anthropological perspective: e.g., Gudeman 1986, 2001; Hann and Hart 2011.

2. Cf. Gudeman 2001.

3. Given that there was no cult or priestess of Athena Parthenos, the Parthenon seems to have been less a temple than an extravagant, temple-shaped votive offering, one that also served as a suitable repository for the "treasures" of imperial Athens. Cf. Kroll 1979; Hurwit 1999, 222–34.

4. Stages and expenses of the project more or less continually monitored and recorded by annual boards of five *epistatai*: Burford 1963, 23–24.

5. Details of the construction of the Parthenon: Stanier 1953; Burford 1963. As many as 2,000 men may have been working on the various temple projects completed in Athens in the period 450–400 BC: Salmon 2001, 204. Major temple projects of the era required a wide range of craftsmen, including architects, carpenters, joiners,

haulers, masons, wax modellers, painters, sawyers, sculptors, under-secretaries, and wood carvers: Randall 1953.

6. Salmon 2001, 199–201.

7. Stanier 1953, 72. Athenian currency denominations were as follows: 6 obols = 1 drachma; 100 drachmai = 1 mna; 6,000 drachmas/60 mnai = 1 talent.

8. Stanier 1953, 73.

9. Documented rates of pay for various workers on the Erechtheion, where carpenters, heavy-load-bearing laborers, and sawyers were all paid one drachma per day: Randall 1953, 208.

10. Stanier 1953, 73.

11. Statue cost was "certainly" somewhere between 600 and 1,000 talents: Stanier 1953, 69, with n.13.

12. Surviving epigraphic records of tribute payments suggest that the average total annual income from this source was somewhere in the 400–500 talent range: Meiggs 1972, 524–59. Higher figure of 600 talents per year: Thucydides 2.13.3.

13. Paying the crews of 60 triremes cost 60 talents of uncoined silver per month: Thucydides 6.8.1.

14. Siege of Samos: Thucydides 1.115–117. The Athenians spent a total of 1,200–1,400 talents on the campaign: Meiggs 1972, 190–93.

15. Assuming again a standard average income of one drachma per day: $(2,000 \times 365) \times 10 = 7,300,000$ dr.

16. Thucydides 2.13.3–5.

17. Overview and analysis: Harris 1995.

18. Harris 1995, 41–61, 65–77, 82–103, 115–200, 206–17.

19. Cf. attempts to ascertain the total financial outlays on the City Dionysia, Great Panathenaia, and other *polis* festivals: Wilson 2008; Pritchard 2015, 27–51.

20. Theopompus of Chios, *Fragments of the Greek Historians* 115 F213.

21. Xenophon, *Constitution of the Athenians* 2.9. In Athens, there were "games and sacrifices throughout the year" (*agosi kai thusiais dietesiais*) as "refreshments for the mind" (*anapaulas tei gnomei*): Thucydides 2.38.1.

22. Rosivach 1994, 9–67.

23. Sacrificial feasts provided by elite households for relatives, friends, and neighbors: e.g., Isaeus 1.31; Menander, *The Bad-Tempered Man* 393–418; Xenophon, *Oeconomicus* 2.4–5; cf. 2.3.11.

24. Rosivach 1994, 14–35. These three calendars can be approximately dated to the 430s (Thorikos), 400–350 (Marathon), and 375–350 (Erchia).

25. Rosivach 1994, 75–76.

26. E.g., Antiphon 6.45; Demosthenes 19.190; Theophrastus, *Characters* 21.11.

27. Discussions of the extant remains of the calendar of Nikomakhos: Lambert 2002; Gawlinski 2007. The "ancestral sacrifices" probably included the Genesia,

Synoikia, Plynteria, and rites for Athena Ergane, Athena Itonia, Heracles, (Ge) Kourotrophos, the Tritopatores, Hyakinthides, Demeter at Eleusis, Apollo Hypo Makrais, Apollo Pythias, and Apollo Prostaterios.

28. *Inscriptiones Graecae* II² 1496.

29. Rosivach 1994, 68–71.

30. Rosivach 1994, 70–73.

31. Xenophon, *Anabasis* 3.2.12.

32. Rosivach 1994, 76–77.

33. Rosivach 1994, 77–83.

34. Cf. Rosivach 1994, 88: "[T]here is no evidence for animals being slain for their meat outside the framework of sacrifice."

35. Rosivach 1994, 95–103.

36. Wilson (2008) and Pritchard (2015, 39) estimate that the total cost of staging the City Dionysia and the Great Panathenaia alone was in the 25–30 talent range each, around half of these outlays coming from "private" sources. The total annual outlay on all *polis* festivals was ca. 100 talents: Pritchard 2015, 49.

37. "Private" items not recorded in the inventories until the early fourth century BC: Harris 1995, 28.

38. Attempt to offer a broad overview of Athenian "public finance:" Pritchard 2015.

39. Kallet 1989.

40. Rosivach 1994, 108–28. The Theoric Fund, which seems to have been originally intended to help cover the costs of staging and attending festivals: Buchanan 1962.

41. Rosivach 1994, 128–42. Deme finances: Whitehead 1986, 149–75.

42. Cf. Rosivach 1994, 143–47.

43. On average around one hundred festival liturgies were performed annually: Pritchard 2015, 43–46.

44. Jim 2014.

45. Jim 2014, 279.

46. Xenophon, *Hellenica* 2.4.14–16.

47. Pindar fr. 75SM. For the translation, see Neer and Kurke 2014, 529. Neer and Kurke go on to make a compelling case that the dance in question would have been staged around the Altar of the Twelve Gods ("navel stone"), which at this point (sometime between 500 and 450) would still have been located in the area of the old "Cecropian agora."

48. Neer and Kurke 2014, 576.

49. A bride's dowry commonly ranged between 5% and 10% of the value of her natal family's estate and helps to explain the observable preference among some wealthier Athenians, like the Bouselidai, for endogamy within their wider *gene*: Cox 1998, 116–20.

50. E.g., Sophocles, *Antigone* 569; Aeschylus, *Eumenides* 658–66.; cf. Aristotle, *Nicomachean Ethics* 1161b27, 1134b10.

51. Dean-Jones 1994, 148–224. Sources for "two seed" theory: e.g., Hippocrates, *On Generation* 6; *On the Nature of the Child* 20; *On Women's Diseases* 1.8; cf. Lloyd 1983, 107n.82.

52. "Male seed only" theory perhaps best explained as a conscious resistance to the prevailing "two seed" theory, which threatened to minimize the father's role in reproduction: Dean-Jones 1994, 149–53.

53. Wilgaux 2011, 221.

54. E.g., Hippocrates, *On Generation* 8.1.

55. E.g., Aristotle, *Magna Moralia* 1202a.

56. E.g., Thucydides 1.139.1.

57. Boutadai: Davies 1971, 348–49. Eumolpidai: Bubelis 2012. Kerykes: Davies 1971, 254–55. Phytalidai and Theseus: Walker 1995, 23. Sophocles and the reception of Asclepius: Clinton 1994.

58. Rutherford 2013, 168–69.

59. History and logistics of classical Athenian cavalry: Kroll 1977; Bugh 1988, esp. 39–78.

60. Complex history of the mechanisms used to fund the Athenian navy: Gabrielsen 2010. While the fourth-century "syntrierarchy" and "symmory" mechanisms were designed to spread this burden among a larger number of families than the original trierarchies had done, these liturgies were still necessarily performed by only a small, affluent portion of the social body. Cf. Davies 1981, 9.

61. Cf. the notorious case of Timotheus, who had to borrow vast sums from the likes of the banker Pasion in the later 370s for this very purpose: Millett 1993.

62. E.g., Aristotle, *Constitution of the Athenians* 27.3; Plutarch, *Life of Cimon* 10, 13.5–7. The "Long Walls" connected the walled Athens with the walled port of Piraeus, forming a defensible urban complex which allowed the Athenians to import goods and access their warships even in times of siege.

63. An apparent mid-fifth-century shift from "private" to "public" sources for the funding of *polis* infrastructural projects: Kallet 2003, 128.

64. Moreno 2007, 322.

65. There were "likely" a mere nine episodes of grain shortage in Athens during the period 508–322, none of which lasted longer than a year: Moreno 2007, 311.

66. Finley 1974. "Advisors" performed an essential function as "teachers" of Demos, especially on complex financial matters: Kallet-Marx 1994.

67. E.g., Ober 1989, 199–202.

68. These were among the virtues most commonly cited in the decrees which honored the benefactors of demes: Whitehead 1986, 241–52.

69. Cf. Finley 1981, 102–103.

70. Ismard 2017.

71. Finley 1981, 111.

72. Attested values of slaves range from around 80 to 90 drachmas for an unskilled female adult to over 400 drachmas for highly skilled male adults: e.g., Demosthenes 27.9; Osborne 2010, 88.

73. Ste. Croix 1981, 113.
74. Plato, *Republic* 578d–579a.
75. Demosthenes' family estate: Demosthenes 27, 28, and 29; Davies 1971, 126–33; Osborne 2010, 95–96.
76. The various members of Lysias' family, their historical fortunes, and their associations with Plato's intellectual circle: e.g., Lysias 12, esp. 4, 19–20; Plato *Republic* 327b; *Phaedrus* 257b.
77. Xenophon, *Oeconomicus* 11.16, cf. 12.2–16.
78. Fortune of 70 talents: Lysias 19.46. Membership of liturgical class: Xenophon, *Oeconomicus* e.g., 7.3,9.20. Further discussion: Davies 1971, 265–68.
79. Hipponikos, Philemonides, and others who leased out slaves for mining: Xenophon, *Ways and Means* 4.15–16. Further evidence for the families in question: Davies 1971, 254–70 (Hipponikos), 403–407 (Nicias), 535 (Philemonides).
80. E.g., Jameson 1978.
81. Cf. Ober 1989, 23–24.
82. History of the treasury of the Hellenotamiai and use of imperial revenues to repay a major loan from Athena: Samons 2000, 70–82, 113–38.

CHAPTER 16

1. Discussion of cases of "bad citizenship" in Athens: Christ 2006.
2. Personal enmities among elites in Athens: Alwine 2015.
3. Plague: Thucydides 2.52–53. Mutilation of Herms and profanation of the Mysteries: Thucydides 6.27.1; Andocides 1. Arginusae and aftermath: Xenophon, *Hellenica* 1.6.1–1.7.35.
4. "Critics of democracy" in Athens: Ober 1998.
5. E.g., Xenophon, *Constitution of the Athenians* 1, 6–8.
6. E.g., Plato, *Gorgias* 483 b–d.
7. E.g., Plato, *Menexenus* 237b–c.
8. Plato, *Republic* 414d–e.
9. Cf. the evolution of a Greek "natural theology": Gerson 1990.
10. References to surviving fragments of texts by Presocratic philosophers follow the conventional numbering system established in H. Diels and W. Kranz, *Die Fragmente der Vorsokratiker* (1903) = DK. Xenophanes on the gods: e.g., DK 21 B11, 14, 15, 23, 25.
11. DK 22 B67.
12. E.g., DK 59 B12. Cf. Cicero, *Academic Questions* 2.118; Lactantius, *Divine Institutions* 1.5.18; Sextus Empiricus, *Against the Mathematicians* 9.6.
13. DK 80 B4.
14. E.g., Cicero, *On the Nature of the Gods* 3.37, 89.
15. Critias fr. 1 Nauck; Euripides fr. 286 Nauck.

16. E.g., Shanske 2007. Thucydides followed as a model or precedent by the nineteenth-century German founders of the modern discipline of history: e.g., Murari Pires 2006; cf. Morley 2014. Thucydides as "the founder of a continuous line of realist thought" in International Relations studies: Monten 2006.

17. Thucydides and tragedy: e.g., Cornford 1907; Lloyd-Jones 1983, 141–44; Greenwood 2006, 83–108.

18. E.g., Rood 1999; Kallett 2001, 97; Hornblower 2004.

19. Finley 1972, 31.

20. Commitment to "precision" (*akribeia*) and use of eye-witness evidence: Thucydides 1.22.1–2.

21. See e.g., Thucydides 1.3–11, 2.15, 6.2.1–3.

22. Schiffman 2011, 5.

23. Schiffman 2011, 57.

24. Thucydides 2.15.2. Cf. Thucydides 2.15.4, where the historian suggests that a number of the older Athenian temples and sanctuaries that were visible in his day must have predated the time of Theseus.

25. Thucydides 1.22.4.

26. Loraux 1980.

27. E.g., Thucydides 2.65.9.

28. Thucydides 1.73–78, 5.86–113.

29. Thucydides 2.15.1.

30. Thucydides 1.2.5.

31. Thucydides 2.36.1.

32. Cf. Jordan 1986, 119–21; Furley 2006, 415–16.

33. E.g., Oost 1975; Jordan 1986.

34. E.g., Jordan 1986, 137–42.

35. E.g., Oost 1975, 192–93; Marinatos 1981.

36. E.g., Thucydides 1.126.5–6, 2.17.2, 2.54, 2.102.5, 3.96, 5.26.3.

37. Kallet 2013. Apollo's promise of help to the Spartans: Thucydides 1.118.3.

38. Cf. Ober 1989, 36–38.

39. The "homogeneous and empty time" of modernity, contrasted with "Messianic time," the continuous present of the "here-and-now" (*Jetztzeit*) experienced by non-modern peoples: Benjamin 1968, 253–64.

CONCLUSION

1. As noted in Chapter 8, a number of anthropologists have already taken some initial steps in this general direction: e.g., Strathern 1992a, 1992b; Graeber 2001, 2011; Descola 2013; Kohn 2013; Rosemont, Jr. 2015.

Bibliography

Acton, P. 2014. *Poiesis: Manufacturing in Classical Athens*. Oxford.

Adcock, F., and D. J. Mosley. 1975. *Diplomacy in Ancient Greece*. London.

Alcock, S. 2002. *Archaeologies of the Greek Past: Landscapes, Monuments, and Memories*. Cambridge.

Aleshire, S. B. 1994. "The demos and the priests: the selection of sacred officials at Athens from Cleisthenes to Augustus." In *Ritual, Politics, Finance: Athenian Democratic Accounts Presented to David Lewis*, edited by R. Osborne and S. Hornblower, 325–37. Oxford.

Allen, D. 2000. *The World of Prometheus: The Politics of Punishing in Democratic Athens*. Princeton.

Alwine, A. T. 2015. *Enmity and Feuding in Classical Athens*. Austin, TX.

Amemiya, T. 2007. *Economy and Economics of Ancient Greece*. Abingdon.

Anderson, G. 2003. *The Athenian Experiment: Building an Imagined Political Community in Ancient Attica*. Ann Arbor, MI.

———. 2007. "Why the Athenians forgot Cleisthenes." In *The Politics of Orality* (Orality and Literacy in Ancient Greece, Vol. 6), edited by C. Cooper, 103–27. Leiden.

———. 2009. "The personality of the Greek state." *Journal of Hellenic Studies* 129: 1–21.

———. 2015. "Retrieving the lost worlds of the past: the case for an ontological turn." *American Historical Review* 120: 787–810.

———. Forthcoming a. "Tyranny and social ontology in classical Athens." In *Tyranny: New Contexts* (Dialogues d'Histoire Ancienne Supplément), edited by S. Lewis. Besançon.

———. Forthcoming b. "Was there any such thing as a non-modern state?" In *State Formations: Global Histories and Cultures of Statehood*, edited by J. L. Brooke, J. C. Strauss, and G. Anderson. Cambridge.

Andreyev, V. N. 1974. "Some aspects of agrarian conditions in Attica in the fifth to third centuries B.C." *Eirene* 12: 5–46.

Asad, T. 1993. *Genealogies of Religion: Discipline and Reasons of Power in Christianity and Islam.* Baltimore, MD.

———. 2003. *Formations of the Secular: Christianity, Islam, Modernity.* Palo Alto, CA.

Bachelard, G. 1984. *The New Scientific Spirit.* Translated by A. Goldhammer. Boston.

Balot, R. 2014. *Courage in the Democratic Polis: Ideology and Critique in Classical Athens.* Oxford.

Barad, K. 1998. "Getting real: technoscientific practices and the materialization of reality." *Differences: A Journal of Feminist Cultural Studies* 10: 87–126.

———. 2007. *Meeting the Universe Halfway: Quantum Physics and the Entanglement of Matter and Meaning.* Durham, NC.

Barker, E. 1951. *Principles of Social and Political Theory.* Oxford.

Barton, C. A., and D. Boyarin. 2016. *Imagine No Religion: How Modern Abstractions Hide Ancient Realities.* New York.

Benjamin, W. 1968. *Illuminations: Essays and Reflections.* New York.

Bennett, T., and P. Joyce. 2010. *Material Powers: Cultural Studies, History and the Material Turn.* London.

Berent, M. 1996. "Hobbes and the 'Greek Tongues.'" *History of Political Thought* 17: 36–59.

———. 2004. "In search of the Greek state: a rejoinder to M. H. Hansen." *Polis* 21: 107–46.

Bers, V., and A. Lanni. 2003. "An introduction to the Athenian legal system." In *Dēmos: Classical Athenian Democracy* (www.stoa.org), edited by C. Blackwell, 1–7.

Bessire, L., and D. Bond. 2014. "Ontological anthropology and the deferral of critique." *American Ethnologist* 41.3: 440–56.

Bhambra, G. 2007. *Rethinking Modernity: Postcolonialism and the Sociological Imagination.* Basingstoke.

———. 2011. "Talking among themselves? Weberian and Marxist historical sociologies as dialogues with 'others.'" *Millennium: Journal of International Studies* 39: 667–81.

Biernacki, R. 2000. "Language and the shift from signs to practices in cultural inquiry." *History and Theory* 39: 289–310.

Bird-David, N. 1990. "The giving environment: another perspective on the economic system of hunter-gatherers." *Current Anthropology* 31: 189–96.

———. 1998. "Beyond 'The Original Affluent Society': a culturalist reformulation." In *Limited Wants, Unlimited Means: A Reader on Hunter-Gatherer Economics and their Environment*, edited by J. M. Gowdy, 115–37. Washington, D.C.

Blanton, R., and L. Fargher. 2008. *Collective Action in the Formation of Pre-Modern States.* West Lafayette, IN.

Bloch, E. 2002. "Hemlock poisoning and the death of Socrates: did Plato tell the truth?" In *The Trial and Execution of Socrates: Sources and Controversies*, edited by T. C. Brickhouse and N. D. Smith, 255–78. New York.

Blok, J. 2009. "Perikles' citizenship law: a new perspective." *Historia* 58: 141–70.

———. 2010. "Deme-accounts and the meaning of *hosios* money in fifth-century Athens." *Mnemosyne* 63: 61–93.

———. 2011. "*Hosiê* and Athenian law from Solon to Lykourgos." In *Clisthène et Lycurgue d'Athènes: autour du politique dans la cité classique*, edited by V. Azoulay and P. Ismard, 233–54. Paris.

———. 2014. "A 'covenant' between gods and men: *hiera kai hosia* and the Greek polis." In *The City in the Classical and the Post-Classical World: Changing Contexts of Power and Identity*, edited by C. Rapp and H. A. Drake, 14–37. New York.

Blundell, S. 1995. *Women in Ancient Greece*. Cambridge, MA.

Blundell, S., and M. Williamson, eds. 1998. *The Sacred and the Feminine in Ancient Greece*. London.

Boedeker, D. 2008. "Family matters: domestic religion in classical Greece." In *Household and Family Religion in Antiquity*, edited by J. Bodel and S. M. Olyan, 229–47. Malden, MA.

Boegehold, A., and A. Scafuro, eds. 1994. *Athenian Identity and Civic Ideology*. Baltimore.

Bohr, N. 1961. *Atomic Physics and Human Knowledge*. New York.

Bordo, S. 1987. *The Flight to Objectivity: Essays on Cartesianism and Culture*. Albany, NY.

———. 1993. *Unbearable Weight: Feminism, Western Culture, and the Body*. Berkeley, CA.

———. 1999. *The Male Body: A Look at Men in Public and in Private*. New York.

Bourriot, F. 1976. *Recherches sur la nature du génos*. Paris.

Brassier, R. 2007. *Nihil Unbound: Enlightenment and Extinction*. London.

Bremmer, J. N. 1994. *Greek Religion* (Greece and Rome New Surveys in the Classics 24). Oxford.

Bresson, A. 2007. *L'économie de la Grèce des cités (fin VIe–Ier siècle): I. Les structures et la production*. Paris.

———. 2008. *L'économie de la Grèce des cités (fin VIe Ier siècle): II. Les espaces de l'échange*. Paris.

———. 2015. *The Making of the Ancient Greek Economy: Institutions, Markets, and Growth in the City-States*. Princeton.

Brock, R. 1994. "The labour of women in classical Athens." *Classical Quarterly* 44: 336–46.

Brown, W. 2010. *Walled States, Waning Sovereignty*. New York.

Bruit Zaidman, L., and P. Schmitt Pantel. 1992. *Religion in the Ancient Greek City*. Cambridge.

Brulé, P. 2003. *Women of Ancient Greece*. Edinburgh.

Bubelis, W. 2012. "Inheritance, priesthoods and succession in classical Athens: the *hierophantai* of the Eumolpidai." In *Families in the Greco-Roman World*, edited by R. Laurence and A. Strömberg, 95–105. London.

Buchanan, J. J. 1962. *Theorika: A Study of Monetary Distributions to the Athenian Citizenry During the Fifth and Fourth Centuries BC.* New York.

Bugh, G. R. 1988. *The Horsemen of Athens.* Princeton.

Burchell, G. et al., eds. 1991. *The Foucault Effect: Studies in Governmentality.* Chicago.

Burford, A. 1963. "The builders of the Parthenon." *Greece and Rome* 10: 23–35.

———. 1993. *Land and Labor in the Greek World.* Baltimore.

Burke, E. M. 1992. "The economy of Athens in the classical era: some adjustments to the primitivist model." *Transactions of the American Philological Association* 122: 199–226.

Burkert, W. 1985. *Greek Religion.* Cambridge, MA.

Burnyeat, M. 1997. "The impiety of Socrates." *Ancient Philosophy* 17: 1–12.

Butler, J. 1990. *Gender Trouble: Feminism and the Subversion of Identity.* New York.

———. 1993. *Bodies that Matter: On the Discursive Limits of "Sex."* New York.

Bynum, C. 1999. "What's all the fuss about the body? A medievalist's perspective." In *History Beyond the Cultural Turn,* edited by V. Bonnell and L. Hunt, 241–80. Berkeley, CA.

Canguilhem, G. 2008. *Knowledge of Life.* New York.

Carawan, E. 2010. "Diadikasiai and the Demotionid problem." *Classical Quarterly* 60: 381–400.

Carneiro, R. 1970. "A theory of the origin of the state." *Science* 169: 733–38.

Cartledge, P. A. 1998. "Laying down *polis* law." *Classical Review* 49: 465–69.

———. 2009. *Ancient Greek Political Thought in Practice.* Cambridge.

Chakrabarty, D. 2000. *Provincializing Europe: Postcolonial Thought and Historical Difference.* Princeton.

———. 2002. *Habitations of Modernity: Essays in the Wake of Subaltern Studies.* Chicago.

Chalmers, D. et al., eds. 2009. *Metametaphysics: New Essays on the Foundations of Ontology.* Oxford.

Chan, J. 2005. "Territorial boundaries and Confucianism." In *Confucian Political Ethics,* edited by D. A. Bell, 113–38. Princeton.

Chan, W., ed. 1963. *A Sourcebook in Chinese Philosophy.* Princeton.

Chartier, R. 1997. *On the Edge of the Cliff: History, Language, and Practices.* Baltimore.

Chrimes, S. B., ed. 1949. *Sir John Fortescue, De Laudibus Legum Angliae.* Cambridge.

Christ, M. R. 2006. *The Bad Citizen in Classical Athens.* New York.

Christesen, P. 2003. "Economic rationalism in fourth-century BC Athens." *Greece and Rome* 50. 1–26.

Clendinnen, I. 1991. *Aztecs: An Interpretation.* Cambridge.

Clinton, K. 1992. *Myth and Cult: The Iconography of the Eleusinian Mysteries.* Stockholm.

———. 1994. "The Epidauria and the arrival of Asclepius in Athens." In *Ancient Greek Cult Practice from the Epigraphical Evidence,* edited by R. Hägg, 17–34. Stockholm.

Cohen, D. 1991. *Law, Sexuality, and Society: The Enforcement of Morals in Classical Athens.* Cambridge.

———. 1995. *Law, Violence, and Community in Classical Athens.* Cambridge.

Cohen, E. E. 1993. *Athenian Economy and Society: A Banking Perspective*. Princeton.

———. 2000. *The Athenian Nation*. Princeton.

Cole, S. G. 2004. *Landscapes, Gender, and Ritual Space: The Ancient Greek Experience*. Berkeley, CA.

Connelly, J. B. 2007. *Portrait of a Priestess: Women and Ritual in Ancient Greece*. Princeton.

Connor, W. R. 1985. "The razing of the house in Greek society." *Transactions of the American Philological Association* 115: 79–102.

———. 1988. "'Sacred' and 'secular': *hiera kai hosia* and the classical Athenian concept of the state." *Ancient Society* 19: 161–88.

Conophagos, E. C. 1980. *Le Laurium antique*. Athens.

Constant, B. 1988. *Constant: Political Writings*, edited and translated by B. Fontana. Cambridge.

Coole, D., and S. Frost, eds. 2010. *New Materialisms: Ontology, Agency, and Politics*. Durham, NC.

Cornford, F. M. 1907. *Thucydides Mythistoricus*. London.

Cox, C. 1998. *Household Interests: Property, Marriage Strategies, and Family Dynamics in Ancient Athens*. Princeton.

———. 2011. "Marriage in ancient Athens." In *A Companion to Families in the Greek and Roman Worlds*, edited by B. Rawson, 231–44. Malden, MA.

Csapo, E., and M. Miller. 1998. "Democracy, empire, and art: toward a politics of time and narrative." In *Democracy, Empire, and the Arts in Fifth-Century Athens*, edited by D. Boedeker and K. Raaflaub, 87–125. Cambridge, MA.

Dagenais, J., and M. R. Greer. 2000. "Decolonizing the Middle Ages: Introduction." In *Decolonizing the Middle Ages* (special issue, *Journal of Medieval and Early Modern Studies* 30), edited by J. Dagenais and M. R. Greer, 431–48.

Darnton, R. 1984. *The Great Cat Massacre and Other Episodes in French Cultural History*. New York.

Daston, L., ed. 2000. *Biographies of Scientific Objects*. Chicago.

Daston, L., and P. Galison. 2007. *Objectivity*. Boston.

Daux, G. 1983. "Le calendrier de Thorikos au Musée J. Paul Getty." *L'Antiquité Classique* 52: 150–74.

Davidson, J. 2011. *Courtesans and Fishcakes: The Consuming Passions of Classical Athens*. Chicago.

Davies, J. K. 1971. *Athenian Propertied Families, 600–300 BC*. Oxford.

———. 1981. *Wealth and the Power of Wealth in Classical Athens*. Salem, NH.

Davies, P., and Gribbin, J. 1992. *The Matter Myth: Dramatic Discoveries that Challenge our Understanding of Physical Reality*. New York.

Davis, N. Z. 1983. *The Return of Martin Guerre*. Cambridge, MA.

Dean-Jones, L. 1994. *Women's Bodies in Classical Greek Science*. Oxford.

Deleuze, G., and F. Guattari. 1987. *A Thousand Plateaus: Capitalism and Schizophrenia*. Minneapolis.

Descola, P. 1994. *In the Society of Nature: A Native Ecology in Amazonia*. Cambridge.

——. 2013. *Beyond Nature and Culture*. Chicago.

Detienne, M. 1989. "The violence of well-born ladies: women in the Thesmophoria." In *The Cuisine of Sacrifice Among the Greeks*, edited by M. Detienne and J.-P. Vernant, 129–47. Chicago.

Deubner, L. 1932. *Attische Feste*. Berlin.

Dillon, M. 2002. *Girls and Women in Classical Greek Religion*. London.

Dodds, E. R. 1951. *The Greeks and the Irrational*. Berkeley, CA.

Dontas, G. 1983. "The true Aglaureion." *Hesperia* 52: 48–63.

Dougherty, C., and L. Kurke, eds. 1998. *Cultural Poetics in Archaic Greece: Cult, Performance, Politics*. New York.

Dover, K. J. 1976. "The freedom of the intellectual in Greek society." *Talanta* 7: 24–54.

Dow, S. 1968. "Six Athenian sacrificial calendars." *Bulletin de Correspondance Hellénique* 92: 170–86.

Dowden, K. 2007. "Olympian gods, Olympian pantheon." In *A Companion to Greek Religion*, edited by D. Ogden, 41–55. Malden, MA.

Easterling, P. E., and J. V. Muir, eds. 1985. *Greek Religion and Society*. Cambridge.

Eley, G. 1996. "Is all the world a text? From social history to the history of society two decades later." In *The Historic Turn in the Human Sciences*, edited by T. J. MacDonald, 193–244. Ann Arbor, MI.

Eley, G., and K. Nield. 2007. *The Future of Class in History: What's Left of the Social?* Ann Arbor, MI.

Engen, D. T. 2010. *Honor and Profit: Athenian Trade Policy and the Economy and Society of Greece, 415–307 BC*. Ann Arbor, MI.

Ermarth, E. D. 2001. "Agency in the discursive condition." *History and Theory* 40: 34–58.

Evans, P. B., Rueschemeyer, D., and Skocpol, T., eds. 1985. *Bringing the State Back In*. Cambridge.

Faraone, C. A. 2008. "Household religion in ancient Greece." In *Household and Family Religion in Antiquity*, edited by J. Bodel and S. M. Olyan, 210–28. Malden, MA.

Farrar, C. 1989. *The Origins of Democratic Thinking: The Invention of Politics in Classical Athens*. Cambridge.

Feyerabend, P. 1975. *Against Method: Outline of an Anarchistic Theory of Knowledge*. Minneapolis.

——. 1987. *Farewell to Reason*. London.

Fingarette, H. 1983. "The music of humanity in the *Conversations* of Confucius." *Journal of Chinese Philosophy* 10: 331–56.

Finley, M. I. 1972. "Introduction." In Thucydides, *History of the Peloponnesian War* (Penguin Classics), translated by R. Warner, 9–32. London.

——. 1973. *The Ancient Economy*. London.

——. 1974. "Athenian demagogues." In *Studies in Ancient Society*, 1–25. London.

——. 1980. *Ancient Slavery and Modern Ideology*. New York.

———. 1981. *Economy and Society in Ancient Greece*, edited with an introduction by B. D. Shaw and R. P. Saller. London.

———. 1983. *Politics in the Ancient World*. Cambridge.

———. 1985. *Democracy Ancient and Modern*. New Brunswick, NJ.

Fisher, N. R. E. 1998. *Slavery in Classical Greece*. Bristol.

Forsdyke, S. 2012. *Slaves Tell Tales: And Other Episodes in the Politics of Popular Culture in Ancient Greece*. Princeton.

Forsen, B. 1996. *Griechische Gliederweihungen: Eine Untersuchung zu ihrer Typologie und ihrer religions- und sozialgeschichtlichen Bedeutung*. Helsinki.

Foucault, M. 1970. *The Order of Things: An Archaeology of the Human Sciences*. New York.

———. 1972. *The Archaeology of Knowledge and the Discourse on Language*. New York.

———. 1977. *Discipline and Punish: The Birth of the Prison*. New York.

———. 1990. *The History of Sexuality: An Introduction, Volume 1*. New York.

———. 2008. *The Birth of Biopolitics. Lectures at the College de France, 1978–79*. Basingstoke.

Fowler, C. 2004. *The Archaeology of Personhood: An Anthropological Approach*. Abingdon.

Foxhall, L. 1989. "Household, gender and property in classical Athens." *Classical Quarterly* 39: 22–44.

———. 1996. "The law and the lady: women and legal proceedings in classical Athens." In *Greek Law in its Political Setting: Justifications not Justice*, edited by L. Foxhall and A. D. E. Lewis, 133–52.

———. 2002. "Access to resources in classical Greece: the egalitarianism of the polis in practice." In *Money, Labour and Land: Approaches to the Economies of Ancient Greece*, edited by P. Cartledge et al., 209–20. London.

———. 2007. *Olive Cultivation in Ancient Greece: Seeking the Ancient Economy*. New York.

Furley, W. D. 2006. "Thucydides and religion." In *Brill's Companion to Thucydides*, edited by A. Rengakos and A. Tsamakis, 415–38. Leiden.

Gabrielsen, V. 2010. *Financing the Athenian Fleet: Public Taxation and Social Relations*. Baltimore.

Gallant, T. W. 1991. *Risk and Survival in Ancient Greece: Reconstructing the Rural Domestic Economy*. Oxford.

Galpin, T. 1983. "The democratic roots of Athenian imperialism in the fifth century BC." *Classical Journal* 79: 100–109.

Gantz, T. 1993. *Early Greek Myth: A Guide to the Literary and Artistic Sources*. Baltimore.

Garlan, Y. 1988. *Slavery in Ancient Greece*. Ithaca, NY.

Garland, R. 1992. *Introducing New Gods: The Politics of Greek Religion*. Ithaca, NY.

Gawlinski, L. 2007. "The Athenian calendar of sacrifices: a new fragment from the Athenian agora." *Hesperia* 76: 37–55.

Geertz, C. 1980. *Negara: The Theatre State in Nineteenth-Century Bali.* Princeton.

Gerson, L. 1990. *God and Greek Philosophy: Studies in the Early History of Natural Theology.* London.

Gierke, O. 1927. *Political Theories of the Middle Age.* Cambridge.

Gill, C. 1996. *Personality in Greek Epic, Tragedy, and Philosophy: The Self in Dialogue.* Oxford.

Ginzburg, C. 1980. *The Cheese and the Worms: The Cosmos of a Sixteenth-Century Miller.* Baltimore.

Goff, B. 2004. *Citizen Bacchae: Women's Ritual Practice in Ancient Greece.* Berkeley, CA.

Gould, J. 1985. "On making sense of Greek religion." In *Greek Religion and Society*, edited by P. E. Easterling and J. V. Muir, 1–33. Cambridge.

Graeber, D. 2001. *Toward an Anthropological Theory of Value: The False Coin of Our Own Dreams.* New York.

———. 2011. *Debt: The First 5,000 Years.* Brooklyn, NY.

———. 2015. "Radical alterity is just another way of saying 'reality': a reply to Eduardo Viveiros de Castro." *HAU: Journal of Ethnographic Theory* 5.2: 1–41.

Greenwood, E. 2006. *Thucydides and the Shaping of History.* London.

Gregory, J. 1991. *Euripides and the Instruction of the Athenians.* Ann Arbor, MI.

Gudeman, S. 1986. *Economics as Culture: Models and Metaphors of Livelihood.* London.

———. 2001. *The Anthropology of Economy: Community, Market, and Culture.* Malden, MA.

Haas, J. 1982. *The Evolution of the Prehistoric State.* New York.

Hadjisteliou-Price, T. 1978. *Kourotrophos: Cults and Representations of the Greek Nursing Deities.* Leiden.

Halbmayer, E. 2012a. "Debating animism, perspectivism and the construction of ontologies." *Indiana* 29: 9–23.

———. 2012b. "Amerindian mereology: animism, analogy, and the multiverse." *Indiana* 29: 103–25.

Hall, E. 1989. *Inventing the Barbarian: Greek Self-Definition through Tragedy.* Oxford.

Hall, J. 1997. *Ethnic Identity in Greek Antiquity.* Cambridge.

Hallowell, A. I. 1960. "Ojibwa ontology, behavior and world view." In *Culture in History: Essays in Honor of Paul Radin*, edited by S. Diamond, 19–52. New York.

Halperin, D. 1990. *One Hundred Years of Homosexuality: And Other Essays on Greek Love.* London.

Hamel, D. 1998. *Athenian Generals: Military Authority in the Classical Period.* Leiden.

Hamilton, A. et al. 1961. *The Federalist Papers.* Edited with an introduction by C. Rossiter. New York.

Hamilton, R. 1984. "Sources for the Athenian Amphidromia." *Greek, Roman, and Byzantine Studies* 25: 243–51.

Hann, C., and K. Hart. 2011. *Economic Anthropology: History, Ethnography, Critique.* Cambridge.

Hansen, M. H. 1987. *The Athenian Assembly in the Age of Demosthenes.* Oxford.

———. 1990. "The political powers of the people's court in fourth-century Athens." In *The Greek City: From Homer to Alexander*, edited by O. Murray and S. Price, 215–43. Oxford.

———. 1991. *The Athenian Democracy in the Age of Demosthenes: Structure, Principles, Ideology*. Oxford.

———. 1996. "The ancient Athenian and the modern liberal view of liberty as a democratic ideal." In *Demokratia: A Conversation on Democracies, Ancient and Modern*, edited by J. Ober and C. Hedrick, 91–104. Princeton.

———. 1998. *Polis and City-State: An Ancient Concept and its Modern Equivalent*. Copenhagen.

———. 2002. "Was the *polis* a state or a stateless society?" In *Even More Studies in the Ancient Greek Polis*, edited by T. H. Nielsen, 17–47. Stuttgart.

———. 2006. *Polis: An Introduction to the Ancient Greek City-State*. Oxford.

Hanson, V. D. 1995. *The Other Greeks: The Family Farm and the Agrarian Roots of Western Civilization*. New York.

———. 1996. "Hoplites into democrats: the changing ideology of Athenian infantry." In *Demokratia: A Conversation on Democracies, Ancient and Modern*, edited by J. Ober and C. Hedrick, 289–312. Princeton.

Haraway, D. 1991. "A cyborg manifesto: science, technology, and socialist-feminism in the late twentieth century." In *Simians, Cyborgs, and Women: The Reinvention of Nature*. New York.

———. 2003. *The Companion Species Manifesto: Dogs, People, and Significant Otherness*. Chicago.

———. 2008. *When Species Meet*. Minneapolis.

Harcourt, B. 2011. *The Illusions of Free Markets: Punishment and the Myth of Natural Order*. Cambridge, MA.

Harman, G. 2002. *Tool-Being: Heidegger and the Metaphysics of Objects*. Chicago.

———. 2010. *Towards Speculative Realism: Essays and Lectures*. Ropley, UK.

Harris, D. 1995. *The Treasures of the Parthenon and the Erechtheion*. Oxford.

Harris, E. M. 2002. "Workshop, marketplace and household: the nature of technical specialization in classical Athens and its influence on economy and society." In *Money, Labour and Land: Approaches to the Economies of Ancient Greece*, edited by P. Cartledge, E. E. Cohen, and L. Foxhall, 67–99. London.

Hedrick, C. W., Jr. 1991. "Phratry shrines of Attica and Athens." *Hesperia* 60: 241–68.

———. 2007. "Religion and society in classical Greece." In *A Companion to Greek Religion*, edited by D. Ogden, 283–96. Malden, MA.

Heisenberg, W. 1958. *Physics and Philosophy: The Revolution in Modern Science*. New York.

Henare, A., M. Holbraad, and S. Wastell. 2007. "Introduction: thinking through things." In *Thinking Through Things: Theorising Artefacts Ethnographically*, edited by A. Henare et al., 1–31. Abingdon, UK.

Henry, M., and S. L. James. 2012. "Women, city, state: theories, ideologies, and concepts in the archaic and classical periods." In *A Companion to Women in the Ancient World*, edited by S. L. James and S. Dillon, 84–95. Malden, MA.

Herington, C. J. 1955. *Athena Parthenos and Athena Polias: A Study in the Religion of Periclean Athens*. Manchester.

Hesse, M. 1966. *Models and Analogies in Science*. South Bend, IN.

———. 1974. *The Structure of Scientific Inference*. Berkeley, CA.

Heywood, P. 2012. "Anthropology and what there is: reflections on 'ontology.'" *Cambridge Anthropology* 30.1: 143–51.

Hobbes, T. 1996. *Leviathan*. Edited with introduction and notes by J. C. A. Gaskin. Oxford.

Hoepfner, W., and E.-L. Schwandner. 1994. *Haus und Stadt im klassischen Griechenland*. 2nd ed. Berlin.

Holbraad, M. 2007. "The power of powder: multiplicity and motion in the divinatory cosmology of Cuban *Ifá* (or *mana*, again). In *Thinking Through Things: Theorising Artefacts Ethnographically*, edited by A. Henare et al., 189–225. Abingdon, UK.

———. 2010. "Ontology is just another word for culture: against the motion (2)." *Critique of Anthropology* 30: 179–85.

———. 2012. *Truth in Motion: The Recursive Anthropology of Cuban Divination*. Chicago.

———. 2013. "Turning a corner: preamble for 'The relative native' by Eduardo Viveiros de Castro." *HAU: Journal of Ethnographic Theory* 3.3: 469–71.

Hopper, R. J. 1968. "The Laurion silver mines: a reconsideration." *Annual of the British School at Athens* 63: 293–326.

Hornblower, S. 2004. *Thucydides and Pindar: Historical Narrative and the World of Epinician Poetry*. Oxford.

Hume, D. 2005. *A Treatise of Human Nature*. Edited with an introduction by M. P. Levine. New York.

Humphreys, S. C. 1978. *Anthropology and the Greeks*. London.

———. 1983. *The Family, Women, and Death: Comparative Studies*. London.

Hunter, V. J. 1989. "Women's authority in classical Greece: the example of Kleoboule and her son." *Echos du Monde Classique* 8: 39–48.

———. 1994. *Policing Athens: Social Control in the Attic Lawsuits, 420–320 BC*. Princeton.

Hurwit, J. 1999. *The Athenian Acropolis: History, Mythology, and Architecture from the Neolithic Era to the Present*. Cambridge.

Ismard, P. 2017. *Democracy's Slaves: A Political History of Ancient Greece*. Cambridge, MA.

James, S. L., and S. Dillon, eds. 2012. *A Companion to Women in the Ancient World*. Malden, MA.

Jameson, M. H. 1978. "Agricultural slavery in classical Athens." *Classical Journal* 73: 122–45.

Jenkins, K., ed. 1997. *The Postmodern History Reader*. London.

Jessop, B. 2008. *State Power: A Strategic-Relational Approach*. Cambridge.

Jiang, Y. 2011. *The Mandate of Heaven and the Great Ming Code*. Seattle.

———. ed. 2014. *The Great Ming Code/Da Ming lü*. Seattle.

Jim, T. S. F. 2014. *Sharing with the Gods: Aparchai and Dekatai in Ancient Greece*. Oxford.

Jones, J. E. 1982. "The Laurion silver mines: a review of recent researches and results." *Greece and Rome* 29: 169–83.

Jones, N. F. 1999. *The Associations of Classical Athens: The Response to Democracy*. New York.

Jordan, B. 1986. "Religion in Thucydides." *Transactions of the American Philological Association* 116: 119–47.

Kahil, L. 1983. "Mythological repertoire of Brauron." In *Ancient Greek Art and Iconography*, edited by W. G. Moon, 231–44. Madison, WI.

Kallet(-Marx), L. 1989. "Did tribute fund the Parthenon?" *Classical Antiquity* 8: 252–66.

———. 1994. "Money talks: rhetor, demos, and the resources of the Athenian empire." In *Ritual, Politics, Finance: Athenian Democratic Accounts Presented to David Lewis*, edited by R. Osborne and S. Hornblower, 227–51. Oxford.

———. 2001. *Money and the Corrosion of Power in Thucydides: The Sicilian Expedition and its Aftermath*. Berkeley.

———. 2003. "*Demos tyrannos*: wealth, power, and economic patronage." In *Popular Tyranny: Sovereignty and its Discontents in Ancient Greece*, edited by K. A. Morgan, 117–53. Austin, TX.

———. 2013. "Thucydides, Apollo, the plague, and the war." *American Journal of Philology* 134: 355–82.

Kant, I. 1983. "Idea for a universal history with a cosmopolitan intent." In *Perpetual Peace and Other Essays, on Politics, History, and Morals*, 29–40. Indianapolis.

Kantorowicz, E. H. 1957. *The King's Two Bodies: A Study in Mediaeval Political Theology*. Princeton.

Katz, M. 1992. "Ideology and 'the status of women' in ancient Greece." *History and Theory* 31: 70–97.

Kearns, E. 1989. *The Heroes of Attica*. (Bulletin of the Institute of Classical Studies Supplement 57). London.

Keller, E. F. 2010. *The Mirage of a Space Between Nature and Nurture*. Durham, NC.

Kennedy, R. F. 2014. *Immigrant Women in Athens: Gender, Ethnicity, and Citizenship in the Classical City*. New York.

Keuls, E. C. 1993. *The Reign of the Phallus: Sexual Politics in Classical Athens*. Berkeley, CA.

Kindt, J. 2012. *Rethinking Greek Religion*. Cambridge.

Kirby, V. 2011. *Quantum Anthropologies: Life at Large*. Durham, NC.

Kohn, E. 2013. *How Forests Think: Toward an Anthropology Beyond the Human*. Berkeley, CA.

———. 2015. "Anthropology of ontologies." *Annual Review of Anthropology* 44.3: 11–27.

Kroll, J. H. 1977. "An archive of the Athenian cavalry." *Hesperia* 46: 83–140.

———. 1979. "The Parthenon frieze as a votive relief." *American Journal of Archaeology* 83: 349–52.

Kuhn, T. 1962. *The Structure of Scientific Revolutions*. Chicago.

Lakoff, G., and M. Johnson. 2003. *Metaphors We Live By*. Chicago.

Lambert, S. D. 1998. *The Phratries of Attica*. 2nd ed. Ann Arbor, MI.

———. 1999. "*IG* II(2) 2345, thiasoi of Herakles and the Salaminioi again." *Zeitschrift für Papyrologie und Epigraphik* 125: 93–136.

———. 2002. "The sacrificial calendar of Athens." *Annual of the British School at Athens* 97: 353–99.

———. 2012. "The social construction of priests and priestesses in Athenian honorific decrees from the fourth century BC to the Augustan period." In *Civic Priests: Cult Personnel in Athens from the Hellenistic Period to Late Antiquity*, edited by M. Horster and A. Klöckner, 67–133. Berlin.

Lambert, S. D., and F. Schuddeboom. 2015. Translation of *Inscriptiones Graecae* II² 1237, available at www.atticinscriptions.com/inscription/RO/5-l-8.

Langdon, M. K. 1976. *A Sanctuary of Zeus on Mount Hymettos*. (Hesperia Supplement XVI.) Princeton.

Lanni, A. 2006. *Law and Justice in the Courts of Classical Athens*. New York.

Larson, J. 2007. "A land full of gods: nature deities in Greek religion." In *A Companion to Greek Religion*, edited by D. Ogden, 56–70. Malden, MA.

Latour, B. 1987. *Science in Action: How to Follow Scientists and Engineers through Society*. Cambridge, MA.

———. 1993. *We Have Never Been Modern*. Cambridge, MA.

———. 2004. *Politics of Nature: How to Bring the Sciences into Democracy*. Cambridge, MA.

———. 2005. *Reassembling the Social: An Introduction to Actor-Network-Theory*. Oxford.

———. 2013. *An Inquiry into Modes of Existence: An Anthropology of the Moderns*. Cambridge, MA.

Latour, B., and Woolgar, S. 1986. *Laboratory Life: The Construction of Scientific Facts*. Princeton.

Lau, D. C., ed. 1979. *Confucius: The Analects*. Harmondsworth.

Le Goff, J. 1980. *Time, Work, and Culture in the Middle Ages*. Chicago.

Lenin, V. I. 1992. *The State and Revolution*. Translated by Robert Service. Harmondsworth.

Lerner, R., and M. S. Mahdi. 1963. *Medieval Political Philosophy: A Sourcebook*. New York.

Lévi-Strauss, C. 1973. *Anthropologie Structurale Deux*. Paris.

Levick, B. 2012. "Women and law." In *A Companion to Women in the Ancient World*, edited by S. L. James and S. Dillon, 96–106. Malden, MA.

Lewis, D. M. 1990. "Public property in the city." In *The Greek City: From Homer to Alexander*, edited by O. Murray and S. Price, 245–63. Oxford.

Lewis, E. K. 1954. *Medieval Political Ideas*, Vol. 1. New York.

Lewis, S. 2002. *The Athenian Woman: An Iconographic Handbook*. London.

L'Homme-Wéry, L.-M. 2000. "La notion de patrie dans la pensée politique de Solon." *Antiquité Classique* 69: 21–41.

Liddell, P. 2007. *Civic Obligation and Individual Liberty in Ancient Athens*. Oxford.

Lienhardt, G. 1985. "Self: public, private. Some African representations." In *The Category of the Person: Anthropology, Philosophy, History*, edited by M. Carrithers et al., 141–55. Cambridge.

Lintott, A. 1982. *Violence, Civil Strife, and Revolution in the Classical City*. London.

Lloyd, G. E. R. 1983. *Science, Folklore and Ideology*. Cambridge.

———. 2011. "Humanity between gods and beasts? Ontologies in question." *Journal of the Royal Anthropological Institute* 17: 829–45.

———. 2012. *Being, Humanity, and Understanding*. Oxford.

Lloyd-Jones, H. 1983. *The Justice of Zeus*. 2nd ed. Berkeley.

Locke, J. 1980. *Second Treatise of Government*, edited by C. B. Macpherson. Indianapolis.

López Austin, A., and L. López Luján. 2001. *Mexico's Indigenous Past*. Norman, OK.

Loraux, N. 1980. "Thucydide n'est pas un collègue." *Quaderni di Storia* 12: 55–81.

———. 1986. *The Invention of Athens: The Funeral Oration in the Classical City*. Cambridge, MA.

———. 1993. *The Children of Athena: Athenian Ideas about Citizenship and the Division between the Sexes*. Princeton.

Lynch, M. "Ontography: investigating the production of things, deflating ontology." *Social Studies of Science* 43.3: 444–62.

Macfarlane, A. 1978. *The Origins of English Individualism*. Oxford.

Mack, W. 2015. *Proxeny and Polis: Institutional Networks in the Ancient Greek World*. Oxford.

MacLachlan, B. 1993. *The Age of Grace: Charis in Early Greek Poetry*. Princeton.

———. 2012. *Women in Ancient Greece: A Sourcebook*. London.

Maitland, F. W. 1901. "The crown as corporation." *Law Quarterly Review* 17: 131–46.

Malin, S. 2001. *Nature Loves to Hide: Quantum Physics and Reality, a Western Perspective*. New York.

Mann, M. 1986. "The autonomous power of the state: its origins, mechanisms, and results." In *States in History*, edited by J. A. Hall, 109–36. Oxford.

Marinatos, N. 1981. "Thucydides and oracles." *Journal of Hellenic Studies* 101: 138–40.

Marriott, M. 1976. "Hindu transactions: diversity without dualism." In *Transaction and Meaning: Directions in the Anthropology of Change and Symbolic Behavior*, edited by B. Kapferer, 109–42. Philadelphia.

———. 1990. "Constructing an Indian ethnosociology." In *India through Hindu Categories*, edited by M. Marriott, 1–39. New Delhi.

Marx, K. 1973. *Grundrisse*. Translated with a foreword by M. Nicolaus. Harmondsworth.

Maurizio, L. 1998. "The Panathenaic procession: Athens' participatory democracy on display?" In D. Boedeker and K. Raaflaub, eds., *Democracy, Empire, and the Arts in Fifth-Century Athens*, 297–317. Cambridge, MA.

McGlew, J. F. 1993. *Tyranny and Political Culture in Ancient Greece*. Ithaca, NY.

Meier, C. 1990. *The Greek Discovery of Politics*. Cambridge, MA.

Meiggs, R. 1972. *The Athenian Empire*. Oxford.

Meillassoux, Q. 2008. *After Finitude: An Essay on the Necessity of Contingency*. New York.

Merkelbach, R. 1972. "Aglauros (Die Religion der Epheben)." *Zeitschrift für Papyrologie und Epigraphik* 9: 277–83.

Mikalson, J. 1977. "Religion in the Attic demes." *American Journal of Philology* 98: 425–35.

———. 1983. *Athenian Popular Religion*. Chapel Hill, NC.

Mill, J. 1967. *Essays on Government, Jurisprudence, Liberty of the Press, and Law of Nations*. New York.

Millett, P. 1993. "Warfare, economy, and democracy in classical Athens." In *Warfare and Society in the Greek World*, edited by J. Rich and G. Shipley, 177–96. London.

Mirowski, P. 1988. *Against Mechanism: Protecting Economics from Science*. Lanham, MD.

Mitchell, T. 1988. *Colonising Egypt*. Berkeley, CA.

———. 1998. "Fixing the economy." *Cultural Studies* 12: 82–101.

———. 1999. "Society, economy, and the state effect." In *State/Culture*, edited by G. Steinmetz, 76–97. Chicago.

———. 2008. "Rethinking economy." *Geoforum* 39: 1116–21.

Monten, J. 2006. "Thucydides and modern realism." *International Studies Quarterly* 50: 3–25.

Moore, J. W. 2015. *Capitalism in the Web of Life: Ecology and the Accumulation of Capital*. London.

Moreno, A. 2007. *Feeding the Democracy: The Athenian Grain Supply in the Fifth and Fourth Centuries BC*. Oxford.

Morley, N. 2014. *Thucydides and the Idea of History*. London.

Morris, I. 1987. *Burial and Ancient Society: The Rise of the Greek City-State*. Cambridge.

———. 1992. *Death Ritual and Social Structure in Classical Antiquity*. Cambridge.

———. 1994. "The Athenian economy twenty years after *The Ancient Economy*." *Classical Philology* 89: 351–66.

———. 1996. "The strong principle of equality and the archaic origins of Greek democracy." In *Demokratia: A Conversation on Democracies, Ancient and Modern*, edited by J. Ober and C. Hedrick, 19–48. Princeton.

———. 2000. *Archaeology as Cultural History: Words and Things in Iron Age Greece*. Oxford.

Morris, I., and B. Powell. 2010. *The Greeks: History, Culture, and Society*. 2nd ed. Upper Saddle River, NJ.

Morton, T. 2007. *Ecology Without Nature: Rethinking Environmental Aesthetics.* Cambridge, MA.

Mossé, C. 1973. *Athens in Decline, 404–86 B.C.* London.

Murari Pires, F. 2006. "Thucydidean modernities: history between science and art." In *Brill's Companion to Thucydides*, edited by A. Rengakos and A. Tsakmakis, 811–38. Leiden.

Nafissi, M. 2004. "Class, embeddedness, and the modernity of ancient Athens." *Comparative Studies in Society and History* 46: 378–410.

———. 2005. *Ancient Athens and Modern Ideology: Value, Theory and Evidence in Historical Sciences.* London.

Nagel, T. 2012. *Mind and Cosmos: Why the Materialist Neo-Darwinian Conception of Nature is almost Certainly False.* New York.

Neer, R. T., and L. Kurke. 2014. "Pindar fr. 75 SM and the politics of Athenian space." *Greek, Roman, and Byzantine Studies* 54: 527–79.

Neils, J., ed. 1992. *Goddess and Polis: The Panathenaic Festival in Ancient Athens.* Princeton.

———. 1999. "Reconfiguring the gods on the Parthenon frieze." *Art Bulletin* 81: 6–20.

Neocleous, M. 2003. *Imagining the State.* Maidenhead, UK.

Nevett, L. 2001. *House and Society in the Ancient Greek World.* Cambridge.

Nietzsche, F. 1976. *Thus Spoke Zarathustra: A Book for Everyone and No One.* Translated by W. Kaufmann. New York.

Nongbri, B. 2008. "Dislodging 'embedded' religion: a brief note on a scholarly trope." *Numen* 55: 440–60.

———. 2013. *Before Religion: A History of a Modern Concept.* New Haven, CT.

Ober, J. 1989. *Mass and Elite in Democratic Athens: Rhetoric, Ideology, and the Power of the People.* Princeton.

———. 1993. "The *polis* as a society: Aristotle, John Rawls and the Athenian social contract." In *The Ancient Greek City-State*, edited by M. H. Hansen, 129–60. Copenhagen.

———. 1996. *The Athenian Revolution: Essays on Ancient Greek Democracy and Political Theory.* Princeton.

———. 1998a. *Political Dissent in Democratic Athens: Intellectual Critics of Popular Rule.* Princeton.

———. 1998b. "Revolution matters: democracy as demotic action (a response to Kurt A. Raaflaub)." In *Democracy 2500? Questions and Challenges*, edited by I. Morris and K. A. Raaflaub, 67–85. Dubuque, IA.

———. 2003. "Tyrant-killing as therapeutic *stasis*: a political debate in images and texts." In *Popular Tyranny: Sovereignty and its Discontents in Ancient Greece*, edited by K. Morgan, 215–50. Austin, TX.

———. 2007. "'I besieged that man': democracy's revolutionary start." In *Origins of Democracy in Ancient Greece*, edited by K. A. Raaflaub, J. Ober, and R. W. Wallace, 83–104. Princeton.

——. 2008. *Democracy and Knowledge: Innovation and Learning in Classical Athens.* Princeton.

——. 2015. *The Rise and Fall of Classical Greece.* Princeton.

Oost, S. I. 1975. "Thucydides and the irrational: sundry passages." *Classical Philology* 70: 186–96.

Osborne, R. 1985. *Demos: The Discovery of Classical Attika.* Cambridge.

——. 2010. "The economics and politics of slavery at Athens." In *Athens and Athenian Democracy*, 85–103. Cambridge.

Ostwald, M. 1986. *From Popular Sovereignty to the Sovereignty of Law: Law, Society, and Politics in Fifth-Century Athens.* Berkeley, CA.

——. 1996. "Shares and rights: 'citizenship' Greek style and American style." In *Demokratia: A Conversation on Democracies, Ancient and Modern*, edited by J. Ober and C. Hedrick, 49–61. Princeton.

Oyama, S. 2000. *Evolution's Eye: A Systems View of the Biology-Culture Divide.* Durham, NC.

Paine, T. 2000. *Thomas Paine: Political Writings.* Cambridge.

Palmer, B. D. 1990. *Descent into Discourse: The Reification of Language and the Writing of Social History.* Philadelphia.

Papazarkadas, N. 2011. *Sacred and Public Land in Ancient Athens.* Oxford.

Parke, H. W. 1977. *Festivals of the Athenians.* London.

Parker, R. 1996. *Athenian Religion: A History.* Oxford.

——. 2005. *Polytheism and Society at Athens.* Oxford.

——. 2011. *On Greek Religion* (Cornell Studies in Classical Philology 60). Ithaca, NY.

Patterson, C. 1981. *Pericles' Citizenship Law of 451/0 BC.* New York.

——. 1986. "*Hai Attikai*: the other Athenians." *Helios* 13: 49–67.

——. 1990. "Those Athenian bastards." *Classical Antiquity* 9: 40–73.

——. 1994. "The case against Neaira and the public ideology of the Athenian family." In *Athenian Identity and Civic Ideology*, edited by A. Boegehold and A. Scafuro, 199–211. Baltimore.

——. 1998. *The Family in Greek History.* Cambridge, MA.

Peels, S. 2014. *Hosios: A Semantic Study of Greek Piety.* Leiden.

Pelling, C. B. R. 2000. *Literary Texts and the Greek Historian.* London.

Petersen, A. 1963. "The philosophy of Nils Bohr." *Bulletin of the Atomic Scientists* 19: 8–14.

Pickering, A. 1981. "The hunting of the quark." *Isis* 72: 216–36.

——. 1984. *Constructing Quarks: A Sociological History of Particle Physics.* Chicago.

——. 1995. *The Mangle of Practice: Time, Agency, and Science.* Chicago.

——. 2016. "The ontological turn: taking different worlds seriously." *Social Analysis* 60: 1–19.

Platt, V. J. 2011. *Facing the Gods: Epiphany and Representation in Graeco-Roman Art, Literature, and Religion.* Cambridge.

Plowden, E. 1816. *Commentaries or Reports*. London.

Prakash, G. 1994. "Subaltern studies as postcolonial criticism." *American Historical Review* 99: 1475–90.

Polanyi, K. 1944. *The Great Transformation: The Political and Economic Origins of our Time*. New York.

Pomeroy, S. B. 1995. *Goddesses, Whores, Wives, and Slaves: Women in Classical Antiquity*. New York.

Pomeroy, S. B. et al. 2008. *Ancient Greece: A Political, Social, and Cultural History*. 2nd ed. New York.

Poovey, M. 1998. *A History of the Modern Fact: Problems of Knowledge in the Sciences of Wealth and Society*. Chicago.

Pritchard, D. 2015. *Public Spending and Democracy in Classical Athens*. Austin, TX.

Raaflaub, K. A. 1996. "Equalities and inequalities in Athenian democracy." In *Demokratia: A Conversation on Democracies, Ancient and Modern*, edited by J. Ober and C. Hedrick, 139–74. Princeton.

———. 1998a. "Power in the hands of the people: foundations of Athenian democracy." In *Democracy 2500? Questions and Challenges*, edited by I. Morris and K. A. Raaflaub, 31–66. Dubuque, IA.

———. 1998b. "The thetes and democracy (a response to Josiah Ober)." In *Democracy 2500? Questions and Challenges*, edited by I. Morris and K. A. Raaflaub, 87–103. Dubuque, IA.

———. 2007. "The breakthrough of *demokratia* in mid-fifth-century Athens." In *Origins of Democracy in Ancient Greece*, edited by K. A. Raaflaub, J. Ober, and R. W. Wallace, 105–54. Princeton.

Randall, R. H. 1953. "The Erechtheum workmen." *American Journal of Archaeology* 57: 199–210.

Rawson, B., ed. 2011. *A Companion to Families in the Greek and Roman Worlds*. Malden, MA.

Rhodes, P. J. 1972. *The Athenian Boule*. Oxford.

Rhodes, P. J., and R. Osborne, eds. 2003. *Greek Historical Inscriptions 404–323 BC*. Oxford.

Roccos, L. J. 1995. "The *kanephoros* and her festival mantle in Greek art." *American Journal of Archaeology* 99: 641–66.

Rood, T. "Thucydides' Persian Wars." In *The Limits of Historiography: Genre and Narrative in Ancient Historical Texts*, edited by C. S. Kraus, 141–68. Leiden.

Rorty, R. 1979. *Philosophy and the Mirror of Nature*. Princeton.

———. 1989. *Contingency, Irony, and Solidarity*. Cambridge.

Rose, N. 1999. *Powers of Freedom: Reframing Political Thought*. Cambridge.

Rosemont, H., Jr., 2005. "Civil society, government, and Confucianism: a commentary." In *Confucian Political Ethics*, edited by D. A. Bell, 46–57. Princeton.

———. 2015. *Against Individualism: A Confucian Rethinking of the Foundations of Morality, Politics, Family, and Religion*. Lanham, MD.

Rosenblum, B., and F. Kuttner. 2011. *Quantum Enigma: Physics Encounters Consciousness*. 2nd ed. New York.

Rosenzweig, R. 2004. *Worshipping Aphrodite: Art and Cult in Classical Athens*. Ann Arbor, MI.

Rosivach, V. J. 1987. "Autochthony and the Athenians." *Classical Quarterly* 37: 294–306.

———. 1994. *The System of Public Sacrifice in Fourth-Century Athens*. Atlanta.

Roussel, D. 1976. *Tribu et cité*. Paris.

Rowbotham, S. 2006. "The trouble with 'patriarchy.'" In *The Feminist History Reader*, edited by S. Morgan, 51–58. London.

Roy, J. 1999. "'Polis' and 'oikos' in classical Athens." *Greece and Rome* 46: 1–18.

Rubel, A. 2014. *Fear and Loathing in Ancient Athens: Religion and Politics During the Peloponnesian War*. London.

Rubinstein, L. 1998. "The Athenian political perception of the *idiotes*." In *Kosmos: Essays in Order, Conflict and Community in Classical Athens*, edited by P. Cartledge et al., 125–43. Cambridge.

Runciman, D. 2000. "Debate: what kind of person is Hobbes's state? A reply to Skinner." *Journal of Political Philosophy* 8.2: 268–78.

Rutherford, I. 2013. *State Pilgrims and Sacred Observers in Ancient Greece: A Study of Theoria and Theoroi*. Cambridge.

Sahlins, M. 1985a. *Islands of History*. Chicago.

———. 1985b. "Hierarchy and humanity in Polynesia." In *Transformations of Polynesian Culture*, edited by A. Hooper and J. Huntsman, 195–223. Auckland.

Salkever, S. G. 1993. "Socrates' Aspasian oration: the play of philosophy and politics in Plato's *Menexenus*." *The American Political Science Review* 87: 133–43.

Sallares, R. 1991. *The Ecology of the Ancient Greek World*. Ithaca, NY.

Salmon, J. 2001. "Temples the measures of men: public building in the Greek economy." In *Economies Beyond Agriculture in the Classical World*, edited by D. J. Mattingly and J. Salmon, 195–208.

Salmond, A. J. M. 2014. "Transforming translations (part 2): addressing ontological alterity." *HAU: Journal of Ethnographic Theory* 4.1: 155–87.

Samons, L. J., II. 2000. *Empire of the Owl: Athenian Imperial Finance*. Stuttgart.

Schiffman, Z. S. 2011. *The Birth of the Past*. Baltimore.

Schmalz, G. C. R. 2006. "The Athenian Prytaneion discovered?" *Hesperia* 75: 33–81.

Schuller, W. 1974. *Die Herrschaft der Athener im ersten attischen Seebund*. Berlin.

Scott, J. W. 1991. "The evidence of experience." *Critical Inquiry* 17: 773–97.

Scott, M. W. 2013. "The anthropology of ontology (religious science?)." *Journal of the Royal Anthropological Institute* 19.4: 859–72.

———. 2014. "To be a wonder: anthropology, cosmology, and alterity." In *Framing Cosmologies: The Anthropology of Worlds*, edited by A. Abramson and M. Holbraad, 31–54. Manchester.

Scullion, S. 2005. "'Pilgrimage' in Greek religion: sacred and secular in the pagan *polis*." In *Pilgrimage in Graeco-Roman and Early Christian Antiquity: Seeing the Gods*, edited by J. Elsner and I. Rutherford, 111–30. Oxford.

Sewell, W. H., Jr. 2005. *The Logics of History: Social Theory and Social Transformation*. Chicago.

Shanske, D. 2007. *Thucydides and the Philosophical Origins of History*. Cambridge.

Shellenberger, M., and T. Nordhaus. 2011. *Love Your Monsters: Postenvironmentalism and the Anthropocene*. Oakland, CA.

Siewert, P. 1977. "The ephebic oath in fifth-century Athens." *Journal of Hellenic Studies* 97: 102–11.

Sinclair, R. K. 1988. *Democracy and Participation in Athens*. Cambridge.

Smith, A. 1976. *The Theory of Moral Sentiments, Volume I*. Oxford.

———. 1986. *The Wealth of Nations, Books I–III*. Harmondsworth.

Smith, A. C. 2003. "Athenian political art from the fifth and fourth centuries: images of political personifications." *Demos: Classical Athenian Democracy* (www.stoa.org), edited by C. Blackwell, 14–23.

Sourvinou-Inwood, C. 1990. "What is *polis* religion?" In *The Greek City: From Homer to Alexander*, edited by O. Murray and S. Price, 295–322. Oxford.

Spencer, H. 1981. *The Man Versus the State: With Six Essays on Government, Society, and Freedom*. Indianapolis.

Spiegel, G. M. 2005. "Introduction." In *Practicing History: New Directions in Historical Writing after the Linguistic Turn*, edited by G. M. Spiegel, 1–31. New York.

Spivak, G. C. 1988. "Can the subaltern speak?" In *Marxism and the Interpretation of Culture*, edited by G. Nelson and L. Grossberg, 271–313. Urbana, IL.

Stanier, R. S. 1953. "The cost of the Parthenon." *Journal of Hellenic Studies* 73: 68–76.

Stanton, G. R., and P. J. Bicknell. 1987. "Voting in tribal groups in the Athenian assembly." *Greek, Roman, and Byzantine Studies* 28: 51–92.

Ste. Croix, G. E. M. 1954/55. "The character of the Athenian empire." *Historia* 3: 1–41.

———. 1981. *The Class Struggle in the Ancient Greek World: From the Archaic Age to the Arab Conquests*. London.

Stehle, E. 2007. "Thesmophoria and Eleusinian Mysteries: the fascination of women's secret rituals." In *Finding Persephone: Women's Rituals in the Ancient Mediterranean*, edited by M. Parca and A. Tzanetou, 165–85. Bloomington, IN.

———. 2012. "Women and religion in Greece." In *A Companion to Women in the Ancient World*, edited by S. L. James and S. Dillon, 191–203. Malden, MA.

Steinbock, B. 2013. *Social Memory in Athenian Public Discourse: Uses and Meanings of the Past*. Ann Arbor, MI.

Stone, I. F. 1989. *The Trial of Socrates*. New York.

Strathern, M. 1988. *The Gender of the Gift: Problems with Women and Problems with Society in Melanesia*. Berkeley, CA.

———. 1992a. *After Nature: English Kinship in the Late Twentieth Century*. Cambridge.

———. 1992b. *Reproducing the Future: Essays on Anthropology, Kinship, and the New Reproductive Technologies*. Manchester.

Strauss, B. S. 1996. "The Athenian trireme: school of democracy." In *Demokratia: A Conversation on Democracies, Ancient and Modern*, edited by J. Ober and C. Hedrick, 313–25. Princeton.

Symes, C. 2011. "When we talk about modernity." *American Historical Review* 116: 715–26.

Thomas, R. 1989. *Oral Tradition and Written Record in Classical Athens*. Cambridge.

Thompson, H. A., and R. E. Wycherley. 1972. *The Athenian Agora. Vol. XIV. The Agora of Athens: The History, Shape, and Uses of an Ancient City Center*. Princeton.

Tilly, C. 1992. *Coercion, Capital, and European States, AD 990–1992*. Cambridge, MA.

Todd, S. C. 1993. *The Shape of Athenian Law*. Oxford.

Toews, J. E. 1987. "Intellectual history after the linguistic turn: the autonomy of meaning and the irreducibility of experience." *American Historical Review* 92: 879–907.

Traill, J. S. 1986. *Demos and Trittys: Epigraphical and Topographical Studies in the Organization of Attica*. Toronto.

Travlos, J. 1971. *Pictorial Dictionary of Ancient Athens*. London.

Tylor, E. B. 1871. *Primitive Culture: Researches into the Development of Mythology, Philosophy, Religion, Language, Art, and Custom. Volume One*. London

van Wees, H. 2004. *Greek Warfare: Myths and Realities*. London.

Vattimo, G. 2011. *A Farewell to Truth*. New York.

Vernant, J.-P. 1980. *Myth and Society in Ancient Greece*. London.

Veyne, P. 1988. *Did the Greeks Believe in Their Myths? An Essay on the Constitutive Imagination*. Chicago.

Vidal-Naquet, P. 1986. *The Black Hunter: Forms of Thought and Forms of Society in the Greek World*. Baltimore.

Viveiros de Castro, E. 1998. "Cosmological deixis and Amerindian perspectivism." *Journal of the Royal Anthropological Institute* 4.3: 469–88.

———. 2003. *AND*. Manchester Papers in Social Anthropology 7.

———. 2014. *Cannibal Metaphysics: For a Post-Structural Anthropology*. Minneapolis.

———. 2015. "Who is afraid of the ontological wolf? Some comments on an ongoing anthropological debate." *Cambridge Journal of Anthropology* 33.1: 2–17.

Vlassopoulos, K. 2007. *Unthinking the Greek Polis: Ancient Greek History Beyond Eurocentrism*. Cambridge.

von Reden, S. 1995. *Exchange in Ancient Greece*. London.

Walbank, M. B. 1978. *Athenian Proxenies of the Fifth Century BC*. Toronto.

Walker, H. 1995. *Theseus and Athens*. New York.

Wang, A. 2000. *Cosmology and Political Culture in Early China*. Cambridge.

Weinstein, B. 2005. "History without a cause? Grand narratives, world history, and the postcolonial dilemma." *International Review of Social History* 50: 71–93.

Wendt, A. 2015. *Quantum Mind and Social Science: Unifying Physical and Social Ontology*. Cambridge.

West, M. L., ed. 1978. *Hesiod's Works and Days*. Oxford.

Whitehead, D. 1977. *The Ideology of the Athenian Metic*. Cambridge.

———. 1986. *The Demes of Attica, 508/7–ca. 250 BC: A Political and Social Study*. Princeton.

Wilgaux, J. 2011. "Consubstantiality, incest, and kinship in ancient Greece." In *A Companion to Families in the Greek and Roman Worlds*, edited by B. Rawson, 217–30. Malden, MA.

Will, E. 1972. *Le monde grecque et l'orient: le Ve siècle (510–403)*. Paris.

Wilson, E. 2007. *The Death of Socrates: Profiles in History*. Cambridge, MA.

Wilson, P. 2000. *The Athenian Institution of the Khoregia: The Chorus, the City and the Stage*. Cambridge.

———. 2008. "Costing the Dionysia." In *Performance, Reception, Iconography: Studies in Honour of Oliver Taplin*, edited by M. Revermann and P. Wilson, 88–127. Oxford.

Winch, P. 1958. *The Idea of a Social Science and its Relation to Philosophy*. London.

———. 1964. "Understanding a primitive society." *American Philosophical Quarterly* 1: 307–24.

Wycherley, R. E. 1957. *The Athenian Agora. Vol. 3. Literary and Epigraphic Testimonia*. Princeton.

Zeitlin, F., and J. Winkler, eds. 1990. *Before Sexuality: The Construction of Erotic Experience in the Ancient Greek World*. Princeton.

Index

Note: Page numbers followed by n and a number indicate endnotes.

The Redoubtable
Mrs. Smith

The Redoubtable Mrs. Smith

a trade union novel

Leif Mills

To order additional copies of this book, contact:
Xlibris Corporation
0-800-644-6988
www.xlibrispublishing.co.uk
orders@xlibrispublishing.co.uk
300178